Rethinking Water Sustainability With Indigenous Land-Based Knowledge and Practice

Ranjan Datta
https://orcid.org/0000-0001-7511-6583
Mount Royal University, Canada

Kevin Lewis
University of Saskatchewan, Canada

Margot Hurlbert
University of Regina, Canada

Jebunnessa Chapola
University of Calgary, Canada

Michelle Whitstone
https://orcid.org/0009-0009-3868-6777
Diné College, USA

John Acharibasam
University of the Fraser Valley, Canada

Vice President of Editorial	Melissa Wagner
Director of Acquisitions	Mikaela Felty
Director of Book Development	Jocelynn Hessler
Production Manager	Mike Brehm
Cover Design	Jose Rosado

Published in the United States of America by
IGI Global Scientific Publishing
701 East Chocolate Avenue
Hershey, PA, 17033, USA
Tel: 717-533-8845 | Fax: 717-533-7115
Website: https://www.igi-global.com E-mail: cust@igi-global.com

Copyright © 2026 by IGI Global Scientific Publishing. All rights reserved. No part of this publication may be reproduced, stored or distributed in any form or by any means, electronic or mechanical, including photocopying, without written permission from the publisher. Use of this publication to train generative artificial intelligence (AI) technologies is expressly prohibited. The publisher reserves all rights to license its use for generative AI training and machine learning model development.

Product or company names used in this set are for identification purposes only. Inclusion of the names of the products or companies does not indicate a claim of ownership by IGI Global Scientific Publishing of the trademark or registered trademark.

Library of Congress Cataloging-in-Publication Data

LCCN: 2026004236 (CIP Data Pending)
ISBN13: 979-8-3373-7559-5
Isbn13Softcover: 979-8-3373-7560-1
EISBN13: 979-8-3373-7561-8
British Cataloguing in Publication Data
A Cataloguing in Publication record for this book is available from the British Library.

All work contributed to this book is new, previously-unpublished material.
The views expressed in this book are those of the authors, but not necessarily of the publisher.
This book contains information sourced from authentic and highly regarded references, with reasonable efforts made to ensure the reliability of the data and information presented. The authors, editors, and publisher believe the information in this book to be accurate and true as of the date of publication. Every effort has been made to trace and credit the copyright holders of all materials included. However, the authors, editors, and publisher cannot assume responsibility for the validity of all materials or the consequences of their use. Should any copyright material be found unacknowledged, please inform the publisher so that corrections may be made in future reprints.

Table of Contents

Foreword v

Preface xi

Chapter 1
Rethinking Water Sustainability Through Indigenous Land-Based Science 1

Chapter 2
Indigenous Traditional Water Knowledge and Practice as Environmental Sustainability 23

Chapter 3
Indigenous Water as Life 43

Chapter 4
Indigenous Water Stories: Self-Determination 61

Chapter 5
Indigenous Language as Water Governance 85

Chapter 6
Indigenous Water as Healing and Resilience 105

Chapter 7
Healing and Resilience: Stories of Land-Based Connections 125

Chapter 8
Protecting Water as Responsibility for Moving Forward 139

Chapter 9
Relational Water Governance in Practice: As Sustainable Health Well Being .. 161

Chapter 10
Indigenous Futures: Ceremony, Climate Resilience, and Water Sustainability
Beyond Crisis 187

Land-Based Camps as a Transformative Method 209

Glossary 219

Compilation of References 223

About the Authors 231

Index 235

Foreword

When I first learned about the work that would eventually become *Rethinking Water Sustainability With Indigenous Land-Based Knowledge and Practice*, I felt an immediate sense of recognition and connection. The themes explored in this book—relationships with land, water, language, ceremony, and community—resonate deeply with the teachings that have guided my own journey as a nêhiyaw (Cree) woman, educator, and scholar. These teachings are not abstract theoretical ideas or philosophical frameworks that exist only in academic texts. They are lived realities carried through language, relationships, responsibilities, and everyday practices that connect us to the land and to future generations.

In many Indigenous worldviews, knowledge is inseparable from relationships. What we know and how we come to know are grounded in our responsibilities to the land, to water, to our communities, and to the ancestors who have entrusted us with these teachings. This book reflects those responsibilities in powerful ways. It reminds readers that Indigenous knowledge systems are not symbolic or metaphorical descriptions of the natural world; rather, they are living systems of governance and responsibility that shape how we live, how we learn, and how we care for the world around us.

As I read the chapters that follow, I was reminded of the urgency of reconceptualizing our relationships with water. Across many parts of the world today, water systems are under increasing pressure from industrial development, climate change, pollution, and unsustainable patterns of resource extraction. Yet within many dominant policy frameworks, water continues to be framed primarily as a resource—something to be managed, controlled, and allocated through technical and economic systems. Indigenous knowledge systems offer a fundamentally different understanding.

For Indigenous peoples, water is not simply a resource. Water is life.

Water is a relative, a living presence within the broader network of relationships that sustain life. In Cree teachings, water—*nipiy*—is understood as sacred and life-giving. It sustains not only our physical bodies but also our spiritual and cultural relationships with the land. To speak about water therefore requires humility, respect, and responsibility. Protecting water is not simply an environmental issue; it is also a moral, cultural, and spiritual responsibility that extends across generations.

This book reminds us that sustainability cannot be achieved solely through technical solutions or policy reforms. Sustainable relationships with land and water require a deeper transformation in how we understand our responsibilities to the natural world. Indigenous knowledge systems have carried such teachings for generations. They offer ways of living that emphasize balance, reciprocity, and respect for the interconnected web of life.

I have known Dr. Kevin Lewis and Dr. Ranjan Datta for many years through our shared involvement in Indigenous education, decolonial research, land-based learning, and community-engaged research. Over the years, our paths have crossed in many spaces where conversations about Indigenous knowledge, governance, language revitalization, and environmental stewardship are taking place. Through these encounters, I have come to appreciate the deep commitments both of these scholars bring to their work and to the communities with whom they collaborate.

Dr. Kevin Lewis's dedication to Cree language revitalization and cultural education has been evident for decades. His work reflects a profound understanding that language is not merely a tool for communication; it is a way of understanding the world. Language carries the teachings, values, and relationships that shape Indigenous worldviews. Through his leadership in land-based cultural education initiatives, Kevin has worked tirelessly to ensure that Cree language and cultural knowledge continue to be transmitted across generations.

Dr. Ranjan Datta's work similarly reflects a strong commitment to Indigenous land-based governance and environmental sustainability. Through his long-standing collaborations with Indigenous communities in Saskatchewan and beyond, he has engaged in community-based research that centers relational accountability and Indigenous knowledge systems. His research emphasizes humility, reciprocity, and respect—values that are essential for building meaningful partnerships between universities and communities.

In many conversations with Kevin and Ranjan over the years, we have spoken about the importance of restoring relationships between land, language, and learning. These conversations reflect broader movements within Indigenous education and research that seek to reconnect knowledge systems that were disrupted by colonial policies and educational institutions.

Kevin once introduced me to the culture camps that form the heart of this book's story. From the moment I learned about these camps, I understood that they represented something much deeper than a program or initiative. They are living expressions of *nēhiýawi-kiskeyitamowina*—Cree ways of knowing.

Culture camps are spaces where knowledge is shared in ways that cannot easily be replicated in classrooms or textbooks. They are places where learning occurs through relationships, through ceremony, through storytelling, and through the everyday practices that connect people to the land.

At these camps, Elders, Knowledge Keepers, and community members gather with youth to share teachings that have been carried across generations. For many Indigenous peoples who grew up with Elders' teachings, it is understood that knowledge is inseparable from place. Land is not merely a setting where learning occurs; the land itself is the teacher. The waters, forests, animals, winds, and seasons all carry teachings that guide how we live in respectful relationships with the world around us.

When children and youth spend time on the land with Elders, they learn how to listen to these teachings. They learn how to observe the environment, how to care for animals and plants, and how to participate in ceremonies that maintain balance within our communities. These teachings cannot be fully understood through observation alone. They must be experienced through participation and relationship.

The *kâniyâsihk* Culture Camps described in this book represent one such space of relational learning. Over time, these camps have grown into year-round gatherings where community members and visitors come together to reconnect with Cree knowledge systems and land-based practices.

Community members and researchers engage in a wide range of traditional activities during these gatherings. They learn how to harvest plants for medicine, fish and snare animals, tan hides, preserve meat and fish, and construct tools and structures that have long been part of Cree lifeways. They also participate in ceremonies such as the Sun Dance, Sweat Lodge, and other cultural gatherings that hold deep spiritual significance. These activities are not isolated cultural practices. They form part of a broader system of knowledge grounded in relationships. Through these practices, Community members and researchers engage begin to understand how Cree teachings connect language, culture, environment, and spirituality.

One of the most important teachings reflected in this book is the concept of *wâhkôhtowin*. This concept describes the network of relationships that connects all beings—human and more-than-human. It reminds us that we do not exist separately from the land or water. Instead, we live within relationships that carry responsibilities.

When people arrive at culture camps, many come from different communities and backgrounds. Some Community members and researchers engage are Indigenous, while others are non-Indigenous individuals who wish to learn about Cree culture and teachings. Yet very quickly, these individuals become part of a shared relational space.

They become family. This is *wâhkôhtowin* in action.

Community members and researchers engage work together to prepare food, gather wood, care for children, and support ceremonies. These everyday acts of cooperation create bonds of trust and responsibility. Through these experiences, people learn values of humility, generosity, and respect for one another.

In many ways, this book captures the spirit of those gatherings. It documents a three-year journey during which scholars, Elders, Knowledge Keepers, youth, and community members gathered on the land to share teachings about water, sustainability, and relational responsibility. These gatherings were not simply research activities. They were learning journeys grounded in ceremony, conversation, and community relationships. Knowledge emerged through shared experiences rather than through formal instruction alone.

As I read this book, I was reminded of many culture camps I have attended throughout my life and career. These gatherings have always been powerful spaces for the transmission of Indigenous knowledge across generations. Elders share teachings about language, land, history, and ceremony with youth who are eager to reconnect with their identities.

Culture camps also play an important role in healing the disruptions caused by colonization and residential schools. Colonial policies sought to break the relationships between language, land, and cultural identity. For many Indigenous communities, these disruptions created deep social and cultural wounds that continue to be felt today.

Culture camps help restore those relationships.

When youth return to the land through these gatherings, they begin to reclaim both language and identity. They develop a deeper understanding of their responsibilities to the land and to their communities. This process strengthens resilience within communities and supports the renewal of Indigenous knowledge systems.

My own research on *nēhiýawi-kiskeyitamowina* has explored how culture camps function as powerful sites of knowledge transmission. At these camps, Elders share teachings about plants, animals, ceremonies, and social responsibilities. Youth learn not only practical skills but also ethical teachings about how to live in respectful relationships with all living beings.

These teachings are especially important today as Indigenous communities confront growing environmental challenges. Across Canada and many other parts of the world, water systems have been disrupted by industrial development, climate change, and colonial governance systems that often ignored Indigenous knowledge and stewardship practices.

Many Indigenous communities continue to face water insecurity and environmental degradation. Boil-water advisories, contaminated watersheds, and declining ecosystems are reminders that the relationship between human societies and water has become deeply imbalanced.

This book approaches these challenges from a perspective grounded in Indigenous teachings. Rather than viewing water as something to be extracted or managed, the authors emphasize that water must be understood as a living relative deserving of respect and protection.

Through stories shared by Elders, community leaders, and Community members and researchers engage in the culture camps, readers are invited to reconsider what sustainability means. Sustainability is not simply about environmental management or technological innovation. It is about relationships—relationships between people, land, water, language, and spirit.

Another important strength of this book lies in its collaborative approach. The authors bring diverse perspectives and experiences to the work. Kevin Lewis contributes deep knowledge of Cree language revitalization and land-based cultural education. Dr. Datta contributes extensive experience in community-engaged research focused on Indigenous environmental governance. Their collaborators bring additional insights from policy research, education, and community engagement.

This collaborative approach reflects a fundamental principle within Indigenous research traditions: knowledge must be shared respectfully and relationally. Research should never be extractive. Instead, it should strengthen communities and contribute to collective learning.

An equally important aspect of this book is its focus on youth. At the culture camps, young people play active roles as learners, helpers, and future leaders. They assist Elders during ceremonies, gather medicines, prepare food, and participate in storytelling circles.

Through these experiences, youth develop a sense of responsibility that extends beyond the camp. They begin to see themselves as caretakers of the land and water. These young people represent the future of Indigenous governance and environmental stewardship.

Reconnecting with their languages and cultural teachings, they gain the tools needed to protect their lands and waters for generations to come. As a land-based educator, I am encouraged by how this book highlights land-based pedagogy as an essential part of Indigenous education. Across many universities today, educators are working to create learning spaces that recognize Indigenous knowledge systems as legitimate and vital forms of scholarship.

Land-based learning challenges conventional educational structures. Instead of separating knowledge from place, it brings learners into direct relationship with land, community, and Elders. It emphasizes experiential learning, relational accountability, and community engagement.

Books such as this one contribute significantly to these broader educational movements. They demonstrate how land-based knowledge systems can inform both community practice and academic research.

The stories shared throughout this book remind us that knowledge is relational and communal. No one learns alone. Elders, families, communities, and the land itself all participate in the process of learning.

As readers move through the chapters, they will encounter teachings about ceremony, governance, resilience, and responsibility. They will also encounter stories of healing—stories of communities reclaiming their relationships with water, land, and culture.

These stories demonstrate that Indigenous knowledge systems remain vibrant and adaptive. Despite centuries of colonial disruption, Indigenous communities continue to carry teachings that can guide more sustainable futures.

The *kâniyâsihk* Culture Camps illustrate how these teachings come alive in practice. Through fishing, medicine gathering, hide tanning, woodworking, storytelling, and ceremony, Community members and researchers engage experience what it means to live within relationships of respect and reciprocity.

These practices are not merely cultural traditions. They are forms of governance and education that sustain communities and ecosystems.

For readers who may be unfamiliar with Indigenous land-based learning, this book offers an invitation—an invitation to listen carefully to the teachings shared by Elders and communities, and to reflect on one's own relationships with land and water.

As I reflect on the work presented in this book, I feel deep gratitude for the Elders, Knowledge Keepers, and community members who continue to share their teachings with generosity and patience. Their wisdom reminds us that knowledge is a gift—and like all gifts, it carries responsibilities. I also wish to acknowledge the dedication of the authors who have worked alongside communities to document these teachings with care and respect. Their commitment to relational research and land-based learning reflects values that are deeply aligned with Indigenous ways of knowing.

It is my hope that readers approach this book with humility and openness. The teachings contained within it are not merely academic ideas. They are living teachings grounded in relationships that have sustained Indigenous peoples for generations.

May this book encourage more people to return to the land, to listen to water, to learn from Elders, and to honour the responsibilities we all share toward the future.

In the spirit of *wâhkôhtowin*, may we continue walking together in good ways.

Angelina Weenie
University of Prince Edward Island, Canada & University of Regina, Canada

Preface

INTRODUCTION

This book arose from a strong sense of responsibility, carried over many years of walking with Indigenous communities, Elders, Knowledge Keepers, women, youth, and land-based educators across Northern and Western Canada. It was not conceived as a scholarly project alone, nor as a technical contribution to sustainability debates; rather, it grew out of relationships, ceremonies, and shared grief and hope expressed during times of environmental instability and cultural resurgence. Water, in all its sacredness, has been the constant teacher throughout this journey. Its presence, absence, movement, and memory shaped the questions that guided this work and the responsibilities that now accompany its writing. As Elders often remind us, stories do not belong to individuals—they are carried and shared to strengthen collective responsibilities. This book is offered in that spirit: an expression of solidarity with communities who continue to protect water, land, and all living relations amid ongoing colonial and ecological disruption.

For years, I (Ranjan Datta) have listened to community members speak about their waters—the rivers that once ran clear and fast, the lakes that sustained families for generations, the springs that were sites of ceremony and healing. These stories were not simply environmental accounts; they were expressions of kinship, memory, and governance. Indigenous communities revealed how intimately water is tied to identity, language, and spiritual practice. They also revealed profound grief: the grief of witnessing water becomes polluted by industrial waste, disrupted by hydroelectric dams, or diminished by drought, permafrost melt, and rising temperatures. Yet intertwined with this grief was a remarkable resilience. Communities shared ceremonies, teachings, and commitments that demonstrated not only their refusal to surrender water to colonial exploitation, but also their insistence that water must be cared for through ethical, relational, and ceremonial governance. This book honors those voices and the responsibilities they carry.

We wrote this book because the need to protect water is urgent. Across the polycrisis of climate change, biodiversity loss, colonial dispossession, and cultural disruption, Indigenous communities continue to face disproportionate burdens while sustaining the ethical systems that could guide regenerative futures. The stories shared throughout this work highlight a truth that cannot be ignored: water governance cannot be separated from cultural governance; ceremony cannot be separated from sustainability; and Indigenous law cannot be separated from environmental protection. Water is not a resource. Water is a living being, a relation, and a teacher who calls us back into responsibility. To write about water in any other way would be to reaffirm the colonial logics that fuel environmental destruction. By contrast, this book seeks to write with water—with breath, humility, and relational accountability.

At the same time, this work responds to the silences, exclusions, and distortions within Western environmental policy, sustainability science, and climate discourse. Far too often, Indigenous knowledge is treated as supplementary to technocratic approaches, or as an "alternative perspective" rather than a governing law. Too often, Indigenous communities are included only as stakeholders or consulted after decisions have already been made. Meanwhile, environmental policies continue to perpetuate the very harms they claim to address, approving pipelines, mines, dams, and extractive projects that threaten the lands and waters Indigenous peoples have protected since time immemorial. This book challenges these narratives and systems by grounding its analysis in Indigenous teachings, community leadership, and land-based relationships that exist outside, and often in defiance of, colonial governance. It insists that true sustainability must emerge from Indigenous sovereignty, not from modified versions of the colonial status quo.

In preparing this book, we spent time at cultural camps, water ceremonies, land-based gatherings, and community workshops where Elders shared teachings about responsibility, relationality, and the ethics of care. Youth spoke about their hopes and fears—about losing language, culture, and land-based knowledge, but also about their determination to lead change through ceremony, activism, and land-based education. Women spoke about the spiritual and ceremonial responsibilities they hold as water carriers and protectors, responsibilities that colonial systems have long ignored or undermined. These teachings shaped every chapter of this book, reminding me that writing is not an intellectual exercise but a relational act. It is an offering that must uphold community protocols, honor the stories entrusted to it, and contribute to the collective work of resurgence and protection.

Solidarity is at the heart of this project. Writing this book required learning how to walk alongside communities rather than ahead of them and understanding solidarity as a verb rather than a concept. Solidarity means trusting Indigenous governance and stepping back when necessary. It means challenging extractive research practices and committing to methodologies grounded in reciprocity, ceremony, and long-term relationship-building. It means honoring Indigenous sovereignty not as an aspiration but as an active reality shaping environmental governance. And it means recognizing that protecting water requires systemic transformation—not only within communities but within universities, governments, and society at large.

We wrote this book to contribute to that transformation. Our hope is that it serves as a bridge between academic discourse and community-based action, between policy debates and the lived experiences of Indigenous peoples who continue to defend their waters under conditions of colonial and ecological threat. We hope it offers a vision for environmental futures grounded not in fear or scarcity but in relational abundance—an abundance drawn from kinship, ceremony, and the strength of Indigenous worldviews that understand sustainability as a way of life.

Above all, this book is written Indigenous land-based learning as water science. It is written for the rivers that continue to carve their paths through threatened territories, for the lakes that hold memory, for the springs that sustain life, and for the rain that replenishes the land. It is written for the generations yet to come, who deserve to inherit waters that are alive, healthy, and loved. And it is written in gratitude for the communities who continue to remind the world that water is life—not metaphorically, but spiritually, politically, and ecologically. Their teachings continue to guide this work, and it is my hope that readers will feel called into relationship, responsibility, and solidarity through these pages.

CHAPTER OVERVIEW

Chapter 1: Rethinking Water Sustainability Through Indigenous Land-Based Science

Chapter 1 introduces Indigenous Traditional Knowledge as a land-based ethical, spiritual, and governance system foundational to environmental sustainability. It frames land as a living relative rather than a resource, emphasizing that sustainable futures cannot emerge without Indigenous sovereignty, relational accountability, and cultural continuity. Through teachings grounded in reciprocity, kinship, and treaty responsibility, the chapter challenges extractive colonial environmental management. It argues that Indigenous governance practices—rooted in ceremony, intergenerational learning, and ancestral law—offer sophisticated environmental frameworks more responsive to ecological change than technocratic Western models. The chapter establishes the book's central argument: that environmental sustainability is inseparable from protecting land-based learning systems, revitalizing Indigenous law, and restoring Indigenous jurisdiction over lands and waters disrupted by colonialism. Ultimately, it sets the conceptual foundation for rethinking sustainability.

Chapter 2: Water Sustainability from the Land-Based Perspective

This chapter centers water as a sacred relative, a spiritual presence, and a living teacher. It documents how colonial policies, industrial extraction, and land dispossession have disrupted Indigenous relationships with water, resulting not only in ecological harm but in cultural, emotional, and spiritual trauma. Through land-based pedagogies, Elders' testimony, and community stories, the chapter reframes water crisis as a relational rupture. At the same time, it highlights Indigenous resurgence—water ceremonies, language revitalization, land-based curriculums, and youth leadership—as pathways for restoring sustainable water governance grounded in responsibility, ethics, and identity.

Chapter 3: Water as Life — Indigenous Teachings and Governance for Environmental Healing

Chapter 3 deepens the discussion of water by exploring how water ceremonies shape Indigenous governance and environmental healing. Led especially by women, grandmothers, and youth, water ceremonies function as legal, scientific, spiritual, and political practices. This chapter challenges Western governance models by presenting Indigenous ceremony as a legitimate and sophisticated system of environmental law and observation. Movements such as Keepers of the Water illustrate how intergenerational teachings inform resistance to industrial harm and guide community-led restoration. Healing and governance become inseparable through relational accountability.

Chapter 4: Reconnecting to the Sacred — Land-Based Healing

This chapter examines how Indigenous communities reconnect to land, ceremony, and spirit amidst ecological and cultural disruption. Through teachings about sacred beings, dreams, prophecy, and memory, Chapter 4 highlights how land-based healing restores relationships with ancestors, animals, and the spiritual world. It explores the transformative power of storytelling, mentorship, and cultural revival, especially for youth and urban Indigenous peoples. The chapter argues that ecological healing requires relational healing, and that ceremony is essential in rebuilding connections fractured by colonialism.

Chapter 5: Indigenous Perspectives on Water Sovereignty

Chapter 5 analyzes colonial energy regimes and introduces Indigenous visions for energy sovereignty. Through stories and case studies, it illustrates how Indigenous nations are using renewable energy, land-based governance, and cultural teachings to reclaim autonomy and reduce reliance on harmful extractive industries. The chapter emphasizes matriarchal leadership and youth involvement as critical to imagining energy systems that are regenerative rather than destructive. Energy sovereignty is framed not merely as infrastructure change, but as spiritual, political, and ecological transformation rooted in Indigenous law and relational ethics.

Chapter 6: Indigenous Land-Based Practice as Indigenous Water Governance

This chapter weaves land-based pedagogy with water sovereignty, emphasizing that Indigenous ceremonial ethics are themselves systems of environmental governance. It centers the leadership of women and Knowledge Keepers who uphold water teachings through offerings, songs, and community action. The chapter critiques Western water policy that excludes spiritual and relational knowledge while presenting land-based practice—water walks, spring blessings, medicine gathering—as legitimate governance mechanisms. Through stories of water protectors and land-based educators, the chapter illustrates how governance is lived, embodied, and inseparable from ceremony.

Chapter 7: Healing and Resilience — Stories of Land-Based Connections

Chapter 7 presents narratives of healing and resilience from Elders, youth, and community leaders. Through stories of fishing, medicine gathering, language speaking, and returning to the land, the chapter illustrates how land-based practices rebuild identity, community wellness, and cultural continuity. Land becomes a site of healing for ecological grief, trauma, and disconnection. Healing is shown not merely as a response to crisis but as an expression of cultural resurgence and survivance.

Chapter 8: Protecting Water as Responsibility for Moving Forward

This chapter advances a relational framework for understanding water as a living relative whose governance relies on Indigenous land-based science, ceremony, and reciprocal responsibility. It traces a thematic arc that moves from core principles of relationality to a forward-looking vision of Indigenous resurgence as sustainable water governance. Drawing on teachings from Elders, community-led practices, and influential Indigenous scholarship, the chapter reframes water governance as an ethical and ceremonial practice rather than a technical exercise. It emphasizes that climate change represents a profound relational rupture—disrupting languages, ecological relationships, and intergenerational knowledge systems.

Chapter 9: Relational Water Governance in Practice: As Sustainable Health Well Being

This chapter situates Indigenous water knowledge as a living system of relational governance that actively shapes sustainability practice, health, and collective responsibility. The chapter demonstrates how water is understood not as a resource to be managed, but as a living relative, legal authority, and ethical guide whose wellbeing is inseparable from that of land and people. Through ceremonial practices, daily offerings, seasonal observation, and community stewardship, Indigenous Peoples enact governance that integrates ecological care with emotional, spiritual, and physical wellbeing. The chapter foregrounds land-based survival, resistance to industrial extraction, and responses to water contamination, showing how governance emerges through protection, grief, and intergenerational responsibility. It also explores urban disconnection, childhood water relationships, collective advocacy, and movements such as Keepers of the Water as expressions of lived governance. Overall, Chapter 9 provides a grounded translation of Indigenous water governance as applied sustainability and health practice rather than abstract principle.

Chapter 10: Indigenous Futures: Ceremony, Climate Resilience, and Water Sustainability Beyond Crisis

The last chapter extends the book's argument by centering Indigenous water ceremonies as future-oriented systems of climate resilience and governance. The chapter moves beyond crisis narratives to show how ceremony operates as law-in-action, embedding sustainability within daily life, intergenerational accountability, and ethical responsibility to water. It examines how Indigenous Peoples interpret climate change, water insecurity, and environmental disruption as relational challenges requiring ceremonial response rather than technical control. Water ceremonies, spiritual signs, and collective gatherings are presented as governance mechanisms that guide adaptation, healing, and decision-making. The chapter also addresses the persistence of colonial water regimes and infrastructural inequities, demonstrating how Indigenous governance continues despite exclusion from formal policy systems. Structurally, the chapter links ceremony to broader governance, legal, and planning contexts, illustrating how Indigenous water knowledge offers anticipatory, resilient pathways for sustaining waterscapes and communities into the future.

Chapter 1
Rethinking Water Sustainability Through Indigenous Land-Based Science

ABSTRACT

Chapter 1 introduces Indigenous Traditional Knowledge as a land-based ethical, spiritual, and governance system foundational to environmental sustainability. It frames land as a living relative rather than a resource, emphasizing that sustainable futures cannot emerge without Indigenous sovereignty, relational accountability, and cultural continuity. Through teachings grounded in reciprocity, kinship, and treaty responsibility, the chapter challenges extractive colonial environmental management. It argues that Indigenous governance practices—rooted in ceremony, intergenerational learning, and ancestral law—offer sophisticated environmental frameworks more responsive to ecological change than technocratic Western models. The chapter establishes the book's central argument: that environmental sustainability is inseparable from protecting land-based learning systems, revitalizing Indigenous law, and restoring Indigenous jurisdiction over lands and waters disrupted by colonialism. Ultimately, it sets the conceptual foundation for rethinking sustainability.

INTRODUCTION

In the polycrisis era intensifying ecological disruptions, and the cumulative effects of ongoing colonial dispossession, this book calls for a fundamental reorientation in how water and energy sustainability is conceptualized, governed, and lived within Indigenous homelands. Rather than treating water and energy as inert resources whose value is determined by their extractive potential, this work insists on a worldview grounded in relational accountability—one in which water is understood as a living being, a relative, and a teacher who demands respect, reciprocity, and responsibility. For Cree, Dene, Métis, and many other Indigenous nations across Northern and Western Canada, sustainability is never about optimizing metrics of carbon reduction or achieving technological efficiency; rather, it is a lived ethic shaped through land-based practices, ceremonial governance, generational responsibility, and deeply rooted kinship with all of creation.

The framework shaping this book emerges from longstanding Indigenous teachings that emphasize that cultural, ecological, and spiritual well-being are inseparable. Indigenous Elders and Knowledge Keepers have warned for generations that distancing human life from land-based relationships produces conditions in which ecological systems collapse and cultural memory erodes. Their warnings—once dismissed or minimized within settler-colonial policy frameworks—now resonate with undeniable clarity as communities confront the polycrisis of climate change, socio-political marginalization, biodiversity loss, and intensified extractive pressures. The chapters that follow take seriously the truth that Indigenous communities have carried forward: when humans lose their relational obligations to water, land, and spirit, the balance of the world becomes unsettled.

This book is grounded in several central commitments. First, Indigenous land-based knowledge is not supplementary, peripheral, or symbolic—it must be the foundation of any meaningful strategy for water and energy sustainability. Second, relational accountability, ceremony, and Indigenous legal orders are not cultural "add-ons" but integral governance frameworks capable of guiding environmental decision-making toward equity, justice, and ecological balance. Third, community-led sustainability emerges through practice: from land-water walks organized by Elders to youth-led language revitalization programs, to cultural camps that reconnect community members with teachings about the relationships between water, spirit, and land. These practices illuminate a path toward sustainability that is deeply relational, embodied, and community-led

Throughout this book, we draw on case studies from Indigenous nations across Treaty 6 and surrounding territories, many of whom are undertaking critical work to reclaim water sovereignty, protect land-based cultural knowledge, and restore reciprocal relationships with the more-than-human world. These cases are not presented simply as examples but as teachings in themselves—teachings that challenge colonial assumptions about governance, underscore the limitations of technocratic climate planning, and reveal the transformative potential of Indigenous-led environmental stewardship. As Indigenous and non-Indigenous land-based scholars working alongside each other, we approach this book as a collaborative act of relational accountability. Our aim is not only to intervene in academic discussions but to contribute to the growing movement for Indigenous-led environmental justice, climate adaptation, and sustainable futures rooted in Indigenous worldviews.

Why This Book Now?

We are living in a historical moment shaped by multiple, intersecting crises—environmental, political, cultural, and spiritual—that together produce a polycrisis deeply felt across Indigenous territories. The impacts of settler colonialism and ongoing resource extraction have created structural vulnerabilities that intensify the effects of climate change. Indigenous communities across the northern and western regions of Canada experience water insecurity, forced displacement, language loss, and environmental degradation at disproportionate rates. Climate change is accelerating glacial melt, destabilizing permafrost, lengthening wildfire seasons, altering water flows, and intensifying drought cycles. These ecological disruptions are not isolated phenomena; they reverberate across kinship systems, ceremonial practices, food sovereignty, and the mental, emotional, and spiritual well-being of communities (Whyte, 2017; Houle, 2022).

This book responds to these urgent realities by centering Indigenous knowledges and community-led governance as the foundation for transformative environmental futures. Indigenous knowledge systems are grounded in philosophies of relationality, where responsibilities toward land and water are lived through ethical practice. Scholars and knowledge holders have long emphasized that sustainability requires a return to reciprocal relationships with land—relationships expressed through ceremony, seasonal practices, ethical harvesting, and intergenerational learning (Kimmerer, 2013; Datta, 2018). These practices constitute forms of science, law, and governance developed over thousands of years of living with land and water.

Indigenous communities are not only among those most affected by climate crisis; they are also at the forefront of climate leadership, innovation, and resurgence. From land-based cultural camps to community-led water monitoring programs, Indigenous nations are implementing their own strategies for restoration, resilience, and adaptation (McGregor, 2021; Wildcat, 2009). These strategies emerge not from abstract policy frameworks but from relationships maintained through generations of ceremonial responsibility, storytelling, and land-based learning. The Elders and youth who guide these efforts are not stakeholders but sovereign leaders who hold intimate knowledge of their territories and of the responsibilities embedded in those relations.

The urgency for this book also stems from ongoing structural challenges within Canada's water and energy governance systems. Hydroelectric development, mining, pipelines, urban expansion, and longstanding boil water advisories reveal the extent to which colonial policy frameworks continue to undermine Indigenous sovereignty and stewardship obligations (Simpson, 2014; LaDuke, 2020). These systems—designed around extraction and efficiency rather than relationship and reciprocity—have contributed to the displacement of communities, contamination of water sources, fragmentation of ecosystems, and erosion of ceremonial and linguistic traditions. As many scholars have noted, the loss of language and the disruption of land-based learning are inseparable from environmental decline (McCarty & Nicholas, 2014; Tuck & Yang, 2012). When traditional water teachings are interrupted, the ethical frameworks that sustain ecological balance become threatened.

Yet, alongside these structural harms, Indigenous resurgence movements are reshaping the political and cultural landscape. From Idle No More to grassroots land-defense actions, Indigenous nations are asserting their rights to govern their territories according to their laws, languages, and life-affirming teachings. Movements such as these offer not only resistance but transformative visions of relational sustainability, where governance is grounded in obligations to human and more-than-human kin (Lewis & Datta, 2023; Borrows, 2020). The worldview informing these movements understands water not as an object of human control but as a sacred relative whose wellbeing determines the wellbeing of all living systems (Arsenault et al., 2018).

This book stands within this moment of resurgence and reorientation. It argues that the path toward sustainable water and energy futures lies not in more advanced technologies, but in revitalizing Indigenous legal orders, relational responsibilities, and community-led governance. To move forward, we must ask: What becomes possible when sustainability is shaped by land-based teachings rather than settler-colonial management systems? What transformations emerge when water governance is grounded in ceremony, kinship, and responsibility? How can researchers, communities, and institutions collaborate ethically and respectfully to support Indigenous-led visions for the future?

The chapters that follow offer grounded responses to these questions. They illuminate the profound potential of Indigenous community-led water and energy sustainability and demonstrate how land-based knowledge can guide climate adaptation strategies that are culturally grounded, environmentally responsive, and justice-oriented. They also document the complexities, tensions, and possibilities of working within and against colonial systems of governance, offering a nuanced understanding of what relational sustainability demands.

In centering Indigenous knowledge, this book aims to shift conversations in environmental policy, climate action, and sustainability science toward frameworks that honour Indigenous sovereignty, community leadership, and ceremonial knowledge. We present this work as part of a growing intellectual and community-driven movement seeking environmental futures grounded in Indigenous laws, languages, and relationships. Ultimately, this book argues that addressing the climate crisis requires more than technical solutions—it requires the renewal of ethical, relational governance grounded in the teachings of the land and water themselves.

LAND-BASED THEORETICAL AND METHODOLOGICAL FRAMEWORK

This book developed following and guided by the land-based theoretical framework that centers *Indigenous Traditional Land-Based Knowledge (ITLBK)* as both a philosophy and a methodology for understanding and responding to water and energy sustainability crises. ITLBK is more than a repository of environmental observations; it is a relational worldview that understands land, water, air, and more-than-human beings as living entities bound together through reciprocal responsibilities (Datta, 2015; Kimmerer, 2013; Wilson, 2008). These knowledge systems are deeply land-based, followed by traditional land-based story sharing storytelling Indigenous Elders, Knowledge-keepers and land-based educators, ceremonies, and land-based teaching. ITLBK guides how we understand climate crisis not as an isolated environmental issue, but as a disruption of relationships that are spiritual, cultural, political, and environmental all at once (Simpson, 2014; Datta, 2018).

A guiding principle in this book's framework is *land-based*, a concept developed by Indigenous scholars Bang et al., (2012), Kimmerer (2013), McGregor (2012), Wilson (2020). Our framework also engages *decolonial environmental governance*, which resists settler-colonial paradigms that extract and manage "resources" without consent or cultural consideration. Decolonial governance demands the return of decision-making power to Indigenous nations and recognizes Indigenous law, language, and ceremony as valid and vital sources of environmental regulation (McGregor, 2021; Tuck & Yang, 2012). In this way, Indigenous *sovereignty* is not solely about jurisdiction or legal recognition, but also about *relational accountability*—living in right relationship with the land, ancestors, community, and future generations.

Methodologically, we are committed to *community-based research* that is participatory, reciprocal, and relational. Our research begins with ceremonies and trustful relationships, is guided by Elders and Knowledge Keepers, and is accountable to the communities with whom we work. *Story-sharing*—not data collection—is the primary method for knowledge gathering. Stories carry protocols, ethics, and teachings; they are not just qualitative evidence but acts of memory, instruction, and resistance (Wilson, 2008; Simpson, 2014).

We are also committed to *matriarchal leadership and intergenerational learning*. Women, grandmothers, and aunties are often the carriers of water teachings and stewards of ceremonial knowledge. Youth are not passive recipients but active participants and knowledge makers. Through land-based camps, healing walks, language immersion, and community gatherings, this work affirms that sustainability must begin with relationships—not technologies or policies alone.

This book's methodological foundation is therefore not simply academic; it is ethical, spiritual, and political. It honors Indigenous worldviews, upholds sovereignty, and insists that sustainability is inseparable from Indigenous resurgence and relational justice. Indigenous communities across Western and Northern Canada are confronting a deeply entangled polycrisis—where the effects of climate change, systemic colonial violence, and cultural disruption coalesce to endanger environments and traditional ways of life. These overlapping crises are not isolated events but mutually reinforcing systems of harm that degrade both land and spirit. At the heart of this crisis lies the ongoing breakdown of traditional water and energy systems, which are not merely technological failures but symptoms of a deeper assault on Indigenous sovereignty, environmental knowledge, and intergenerational continuity (Whyte, 2020; McGregor, 2021).

Water and energy infrastructures in these regions have historically been developed without Indigenous consent, designed through settler-colonial logics of extraction and control. Hydroelectric dams, oil sands development, mining operations, and pipeline construction have devastated traditional territories—displacing communities, altering hydrological patterns, and destroying culturally significant water bodies (Baker et al., 2022; Latulippe & Klenk, 2020). These systems are not neutral; they carry the legacy of colonialism, encoded in steel and concrete, severing the relational and ceremonial ties Indigenous peoples have with their lands.

Thus, this book centers the truth that water and energy systems are spiritual and cultural landscapes. In Cree and many other Indigenous traditions, water is a living relative—nipiy—with whom one holds obligations, not ownership. When rivers are dammed, sacred springs are lost, and lakes are polluted, what vanishes is not only environmental stability but language, prayer, memory, and kinship (Borrows, 2020; Craft, 2021). Traditional practices like fishing, gathering at seasonal camps, and conducting water ceremonies—spaces where knowledge is transferred across generations—are undermined by environmental degradation and forced economic dependency.

Currently, over 30 Indigenous communities in Canada remain under long-term boil-water advisories, while many more faces rising energy poverty, lacking access to affordable, reliable, and clean power (Indigenous Services Canada, 2024). These conditions are not accidental—they are systemic. Meanwhile, fossil fuel companies continue to extract energy from Indigenous territories, deepening the climate crisis and exacerbating social, cultural, and ecological harm (Hoicka et al., 2022; Levac et al., 2023).

The cultural consequences are profound. The erosion of land-based practices leads to language loss, spiritual grief, and identity fragmentation, particularly among youth who may be disconnected from Elders, stories, and ceremonies. Yet within this crisis, Indigenous communities are leading the way forward—through cultural regeneration, resistance, and innovation. Land-based camps, youth-led water walks, and language revitalization projects reconnect communities to ancestral knowledge while building ecological resilience.

This chapter calls for a paradigm shift: climate adaptation must not be reduced to technical innovation. It must be rooted in cultural regeneration, community governance, and ceremonial healing. Indigenous peoples are not only disproportionately impacted by the climate crisis—they are also uniquely positioned to lead responses that center kinship, responsibility, and survivance. In these communities, water and energy are not resources—they are relations. Any sustainable future must begin by honoring that truth.

STORY-SHARING AS INDIGENOUS METHOD

This book is focused on Indigenous land-based camp and collective story-sharing as a research methodology and methods rather than following western data-collection technique. Within Indigenous land-based camp, stories are not illustrative anecdotes nor qualitative "evidence" to be extracted, coded, and abstracted from their relational contexts. As Wilson (2008) explains, Indigenous knowledge is relational, meaning that knowledge exists within relationships among people, land, water, ancestors, and future generations. Story-sharing therefore functions as both method and governance practice—an ethical process through which teachings, responsibilities, and laws are transmitted across generations.

Following Kovach (2021), story-sharing is understood here as a dialogical, consent-based, and relational practice rooted in Indigenous epistemologies. Stories emerge through land-based encounters, ceremonial gatherings, water walks, cultural camps, and intergenerational conversations guided by Elders, Knowledge Keepers, and community leaders. These stories are not detached from place; they are anchored in specific waters, territories, and kinship networks. As such, meaning is not derived through decontextualization but through careful listening, reflection, and accountability to the relationships within which stories are shared.

Importantly, story-sharing resists extractive research logics. Stories are shared because relationships exist—not because researchers seek information. Knowledge unfolds at the pace determined by land, ceremony, and community readiness. Silence, repetition, emotion, and spiritual language are not methodological limitations but critical dimensions of Indigenous knowledge transmission. This methodological orientation fundamentally challenges Western research traditions that privilege speed, categorization, and control.

COMMUNITY CONSENT AND RELATIONAL ACCOUNTABILITY

Consent in this research is not a one-time procedural requirement but an ongoing relational responsibility. Community consent is understood as collective, iterative, and embedded within long-term relationships rather than institutional forms alone. Consent is sought through ceremony, dialogue with Elders and community leaders, and shared decision-making about what knowledge may be shared, how it may be interpreted, and where it may circulate. Relational accountability guides every stage of the research process—from entering community spaces, to listening and witnessing stories, to writing and dissemination. This means that the research is accountable not only to academic audiences but, first and foremost, to the communities and lands from which the knowledge emerges. Findings are returned to communities through gatherings, conversations, and community-accessible formats, ensuring that research contributes to cultural continuity, governance capacity, and community-defined priorities rather than external extraction. In this research our Indigenous author Dr. Kevin Lewis. As an experienced land-based scholar led the cultural camp with second Indigenous land-based scholar and educator Michelle Whitstone. Our ethical framework aligns with Indigenous research principles articulated by Wilson (2008), Kovach (2021), and Smith (2012), which emphasize responsibility, reciprocity, and respect as foundational to ethical Indigenous research. Knowledge shared in this book is therefore held with care, humility, and an ongoing commitment to relational integrity.

ANALYTICAL FRAMEWORK: INTERPRETING STORIES AS INDIGENOUS GOVERNANCE KNOWLEDGE

Analytically, stories in this book are not treated as thematic fragments but as relational teachings that illuminate Indigenous governance systems Borrows (2020). Interpretation occurs through a layered process that honours Indigenous epistemologies:

- *Relational focusing* – locating each story within its land, water, and community context.
- *Deep listening* – attending to what is said, what is repeated, and what is left unsaid.
- *Ceremonial meaning* – recognizing that teachings often emerge through metaphor, spirituality, and relational language rather than direct instruction.
- *Governance implications* – understanding stories as expressions of Indigenous law, responsibility, and decision-making frameworks.

Rather than reducing stories to coded themes, this approach foregrounds synthesis, relational coherence, and accountability (Arsenault et al. 2018). Stories are interpreted alongside Indigenous scholarship on water governance, land-based pedagogy, and sovereignty, allowing Indigenous voices to remain central while situating them within broader intellectual conversations.

SCOPE OF THE BOOK

This book is built upon, guided by, and deeply accountable to a land-based theoretical and methodological framework that centers Indigenous Traditional Land-Based Knowledge (ITLBK) as both a philosophical orientation and a methodological pathway for understanding and responding to contemporary water and energy sustainability crises. ITLBK is not a collection of empirical observations about the natural world, nor is it a static archive of cultural practices. Instead, it is a living relational worldview—one that recognizes land, water, air, and more-than-human beings as animate relations bound through reciprocal obligations of care, respect, and responsibility (Datta, 2015; Kimmerer, 2013; Wilson, 2008). Rooted in thousands of years of land-based governance, ceremony, and storytelling, ITLBK teaches that environmental degradation is inseparable from spiritual, cultural, political, and relational disruption. In this sense, climate crisis is not understood simply as an ecological problem but as an expression of broken relationships, a fracturing of the ethical and ceremonial responsibilities that sustain life (Simpson, 2014; Datta, 2018; Whyte, 2020).

A core principle of this framework is being land-based, a concept articulated in the work of Bang et al. (2012), Kimmerer (2013), McGregor (2012), Wilson (2020), and many Indigenous community educators whose teachings foreground how knowledge emerges through active and ongoing relationships with land. Land based thinking shifts climate discourse away from settler-colonial paradigms that treat "resources" as extractable objects, toward frameworks grounded in relational accountability and Indigenous laws. As such, this book also engages with decolonial environmental governance, which challenges the imposed authority of colonial structures that manage water and energy without Indigenous consent, cultural recognition, or relational ethics. Decolonial governance demands the return of decision-making power to Indigenous nations and affirms Indigenous languages, ceremonies, and legal orders as vital and legitimate forms of environmental governance (McGregor, 2021; Tuck & Yang, 2012). Through this lens, sovereignty is not limited to legal jurisdiction—it is the collective responsibility to live in right relationship with land, ancestors, community, and future generations.

Methodologically, this book emerges from community-based, relational, and participatory research practices. Our work begins in ceremony, in the preparation and accountability that come from prayer, protocols, and shared intention. Relationships with Elders, Knowledge Keepers, land-based educators, and community members form the foundation of our process. These relationships guide not only what knowledge is gathered, but how it is gathered, interpreted, and returned to the community. In this framework, story-sharing—rather than data "collection"—is the central method. Stories are vessels of law, memory, ethics, and instruction. They carry protocols and teachings that are not reducible to thematic codes or analytical categories (Wilson, 2008; Simpson, 2014). Story-sharing is therefore a practice of reciprocity, witnessing, and learning, not extraction.

Another core methodological commitment is to matriarchal leadership and intergenerational learning. In many Indigenous nations, women, grandmothers, and aunties hold water teachings and carry ceremonial responsibilities that shape community governance. Their leadership forms a central pillar in the restoration of relationships with water and land. Elders and Knowledge Keepers provide teachings that guide ethical research practices, while youth serve as active knowledge makers who shape the continuity of land-based governance. Through community-led cultural camps, water ceremonies, healing walks, language immersion, and gathering-based teachings, intergenerational learning becomes both a method and a site of resurgence. Sustainability, in this context, is not primarily a technological aspiration but a relational practice cultivated through time, ceremony, and community.

Thus, the methodological foundation of this book is not merely academic—it is relational, ethical, spiritual, and fundamentally political. It affirms Indigenous sovereignty, honours Indigenous worldviews, and insists that environmental sustainability cannot be separated from Indigenous resurgence and relational justice. This book therefore emerges at a moment in which Indigenous communities throughout Western and Northern Canada are confronting a deeply entangled polycrisis—where the accelerating effects of climate change converge with systemic colonial violence and ongoing cultural disruption. In these regions, climate impacts such as melting permafrost, drought, intensified wildfires, unpredictable water flows, and ecosystem degradation are inseparable from the colonial structures that have long undermined Indigenous governance, land-based knowledge systems, and intergenerational continuity (Whyte, 2020; McGregor, 2021; Ford et al. 2020).

Water infrastructures across these territories have historically been developed without Indigenous consent and shaped by extractive logics that prioritize settler interests. Hydroelectric dams, oil sands operations, mining activities, and pipeline networks have permanently altered lands and water systems, causing displacement, contamination, and the destruction of culturally significant sites (Baker et al., 2022; Latulippe & Klenk, 2020). These infrastructures carry with them the material and symbolic legacy of colonialism: they are built upon the dispossession of Indigenous peoples and the suppression of Indigenous laws regulating water, land, and energy relationships. The resulting environmental disruptions undermine ceremonial practices, food sovereignty, and cultural teachings grounded in water and land.

This book therefore centers the understanding—shared across Cree, Dene, Métis, and many Indigenous nations—that water and energy systems are not material commodities but spiritual and cultural landscapes. Water, or *nipiy*, is a living relative, a being with whom humans holds responsibilities rather than rights. When rivers are dammed, springs are contaminated, or lakes are diminished, the consequences are not solely ecological; they reverberate through language, memory, ceremony, and kinship (Borrows, 2020; Craft, 2021). Traditional practices such as fishing, gathering at seasonal camps, and conducting community water ceremonies are forms of governance and knowledge transmission. Their decline through environmental degradation interrupts intergenerational knowledge pathways and contributes to cultural loss and spiritual grief.

In this context, the ongoing water and energy insecurity facing Indigenous communities is not accidental. Today, more than thirty Indigenous communities remain under long-term boil water advisories, while many others face rising energy poverty and unreliable infrastructure (Indigenous Services Canada, 2024). At the same time, industries continue to extract energy from Indigenous territories to fuel the national economy, deepening the ecological and cultural harms of the climate crisis. These structural inequalities illustrate a deeply colonial pattern: Indigenous lands supply the nation with energy and water resources, while Indigenous communities bear the costs of environmental decline, polluted ecosystems, and inadequate infrastructure (Hoicka et al., 2022; Levac et al., 2023).

The cultural consequences of these overlapping crises are profound. As land-based practices are disrupted, language loss accelerates, youth experience disconnection from Elders and ceremonial teachings, and many communities navigate collective grief linked to environmental destruction. Yet, despite these challenges, Indigenous nations are leading transformative responses rooted in cultural resurgence, resistance, and innovation. Land-based camps, youth-led water walks, land-water reclamation movements, and community-driven sustainability initiatives reconnect people to ancestral knowledges while strengthening ecological resilience.

This chapter calls for a paradigm shift: climate adaptation cannot be reduced to technical solutions or policy frameworks designed within colonial institutions. Instead, effective climate response must emerge from cultural regeneration, community governance, relational accountability, and ceremonial healing. Indigenous communities are not only disproportionately impacted by climate change—they also carry the relational philosophies, land-based practices, and governance systems necessary for imagining and enacting sustainable futures. Within these nations, water and energy are not "resources"—they are relations whose care must guide environmental decision-making.

LAND-BASED THEORETICAL AND METHODOLOGICAL FRAMEWORK

This book is grounded in a land-based theoretical and methodological framework that centers Indigenous Traditional Land-Based Knowledge (ITLBK) as a foundational worldview, ethical orientation, and embodied methodology for responding to contemporary water and energy sustainability challenges. ITLBK refuses the fragmentation of knowledge into isolated scientific, cultural, or spiritual domains. Instead, it understands land, water, air, spirit, and all more-than-human beings as living relations with whom humans share reciprocal responsibilities (Datta, 2015; Kimmerer, 2013; Wilson, 2008). These responsibilities emerge from thousands of years of practicing relational governance within homelands, where knowledge is not owned but lived, renewed, shared, and held through story, ceremony, kinship, and intergenerational learning. Thus, ITLBK is not a set of cultural "beliefs" but a sophisticated relational science deeply attuned to ecological patterns, spiritual ethics, and the social conditions necessary for collective survival.

In this framework, climate crisis is not merely an environmental issue but a profound rupture of relational responsibilities. Land-based knowledge teaches that when relationships with land and water are broken—through extraction, pollution, displacement, or political suppression—ecological instability inevitably follows. This instability radiates into cultural, linguistic, ceremonial, emotional, and spiritual realms. For Indigenous Knowledge Keepers, climate change is inseparable from colonial dispossession, jurisdictional fragmentation, and the erosion of Indigenous laws. The crisis is therefore multidimensional, unfolding simultaneously as environmental degradation, cultural interruption, and the undermining of Indigenous self-determination (Simpson, 2014; Datta, 2018). This book is guided by the truth that ecological renewal must be accompanied by relational renewal—an ethical and political restoration of Indigenous responsibilities to the land and waters.

LAND-BASED THEORY: RELATIONAL, CEREMONIAL, AND SOVEREIGN

A central theoretical pillar guiding this work is *land-based relationality*, articulated widely by Indigenous scholars and communities (Bang et al., 2012; Kimmerer, 2013; McGregor, 2012; Wilson, 2020). Land-based relationality teaches that knowledge emerges through living relationships with land—not from abstraction or distance. Land is teacher, relative, memory-holder, and lawmaker. Water is not a commodity but a living being who instructs us in humility, flow, care, and collective responsibility. Energy is not a discrete resource but a manifestation of interdependent systems that support the continuity of life.

Understanding sustainability through this relational framework challenges Western environmental governance, which often prioritizes technical efficiency, economic optimization, and mechanistic management systems. Such frameworks extract knowledge from land rather than learning *with* land. They treat water governance as an administrative task rather than a sacred and legal responsibility. Land-based theory intervenes by insisting that sustainable water and energy futures cannot be designed through frameworks that ignore or undermine Indigenous laws, languages, and ceremonial obligations.

This book is equally grounded in decolonial environmental governance, which critiques the colonial logics embedded in Canada's water and energy systems. These systems often govern Indigenous territories without consent, oversight, cultural recognition, or accountability to Indigenous laws. Decolonial governance does not simply call for inclusion or consultation—it demands the return of decision-making power to Indigenous nations and the restoration of relationships through which environmental authority is traditionally held (Tuck & Yang, 2012; McGregor, 2021). Within this worldview, sovereignty is relational: it is the collective responsibility to uphold the teachings passed down through Elders, ancestors, and the land itself.

ITLBK therefore positions environmental governance not as a neutral technical domain but as a deeply ethical and political sphere. Decisions about water, energy, forests, and land cannot be separated from decisions about the future of Indigenous languages, ceremonies, community health, and intergenerational continuity. Thus, sustainability becomes not a policy objective but a living expression of Indigenous law.

STORY-SHARING AS METHODOLOGY

Methodologically, this book is rooted in community-based, participatory, and relational research practices derived from ITLBK. Research does not begin with problem statements or hypotheses. It begins with ceremony—with offerings, tobacco, prayer, and the collective intention-setting that establishes an ethical foundation for the work. Knowledge gathering is shaped by reciprocal relationships between researchers, Elders, community members, and the land. These relationships determine the pace, direction, and depth of learning, resisting academic pressures for extraction or rapid data production.

The primary method used in this work is story-sharing, a longstanding Indigenous practice through which teachings, histories, laws, and relational ethics are transmitted (Wilson, 2008; Simpson, 2014). Story-sharing differs from Western narrative inquiry: stories are not "data." They are living relationships that carry protocols, responsibilities, and spiritual significance. Elders often share stories in ways that require patience, humility, and the willingness to sit with ambiguity. A story may contain teachings about water governance, kinship responsibilities, community law, or ceremonial ethics—but understanding these teachings requires a relational commitment. Thus, story-sharing is an embodied methodology that involves listening, witnessing, reflecting, and walking alongside communities.

This approach also emphasizes matriarchal leadership and intergenerational learning. Water teachings are often carried by women, grandmothers, and aunties who hold ceremonial roles essential to the continuity of community governance. Their knowledge is not symbolic but foundational to how water and energy relationships are understood and practiced. Youth also hold central roles as knowledge carriers, activists, ceremony helpers, land stewards, and language learners. Through land-based camps, water walks, and cultural gatherings, youth participate in relational governance, learning from Elders while shaping the future of environmental leadership.

POSITIONALITY

This book is written by Indigenous and non-Indigenous scholars working in relational accountability with Indigenous communities across Western and Northern Canada last 6 years. Our positionalities are not neutral. They are shaped by our identities, responsibilities, and long-term relationships with land and community. As land-based researchers, educators, and community collaborators, we approach this work with humility, recognizing that Indigenous knowledge is not owned by researchers but entrusted through relationships. Our role is not to speak for communities, but to walk alongside them—listening, learning, and amplifying Indigenous governance teachings as they are shared.

This work is therefore both scholarly and relational. It emerges from years of community-based engagement, ceremonial participation, and accountability to Elders, Knowledge Keepers, women, and youth who carry water teachings. The responsibility for any shortcomings in interpretation rests with us as authors.

ETHICS OF RELATIONAL ACCOUNTABILITY

The methodological foundation of this book is not solely academic—it is ethical and ceremonial. Knowledge shared by communities is held with respect, accountability, and reciprocity. This means that research outcomes must be returned to communities in ways that support ongoing cultural, environmental, and political work. Relational accountability ensures that the research does not reproduce extractive practices. Instead, it strengthens community relationships, supports resurgence, and affirms Indigenous sovereignty.

Research undertaken within this framework must continually ask: *Who benefits? Who is accountable? Who is responsible for ensuring that this knowledge is honored, protected, and enacted appropriately?* These questions ground the work in Indigenous ethical traditions rather than academic conventions.

HOW TO USE THIS BOOK

This book is written for multiple audiences, each of whom may enter the text from different pathways:

- *Policymakers and practitioners* may focus on *Chapters 1, 4, and 8,* which address governance frameworks, structural challenges, and forward-looking recommendations.
- *Researchers and students* may engage with the full manuscript, with particular attention to *Chapters 1, 5, and 8,* which elaborate theoretical, methodological, and policy dimensions.
- *Community members, educators, and land-based practitioners* may find *Chapters 3, 6, and 7* most relevant, as these chapters center ceremony, land-based practice, healing, and intergenerational learning.

Readers are invited to move through the book non-linearly, returning to chapters as relationships, responsibilities, and questions evolve.

RESEARCH GAPS AND CONTRIBUTIONS

Despite growing recognition of Indigenous knowledge within environmental and sustainability scholarship, significant gaps remain in how water and energy governance are conceptualized, practiced, and institutionalized within policy, research, and planning frameworks. This book responds directly to three interrelated gaps and advances corresponding contributions that are taken up and operationalized through the policy pathways articulated in Chapter 8. First, there is a persistent lack of Indigenous-led water and energy governance frameworks within policy and decision-making systems. Indigenous communities are frequently positioned as "stakeholders" or "rights holders," yet dominant governance regimes continue to marginalize Indigenous jurisdiction, legal orders, and decision-making authority over water and energy systems. Existing policy approaches prioritize technocratic management, market efficiency, and state or corporate control, leaving limited space for Indigenous governance grounded in land-based responsibility. This book addresses this gap by centering Indigenous sovereignty and self-determination as foundational to sustainability governance, offering community-led models that inform the policy recommendations advanced in Chapter 8, particularly those directed at governments, regulatory institutions, and funding agencies.

Second, ceremonial and land-based knowledge remain marginalized within sustainability science and environmental policy. While Indigenous knowledge is increasingly acknowledged, it is often treated as cultural context rather than as a rigorous system of environmental governance, monitoring, and law. Ceremonies, water teachings, and land-based practices are rarely recognized as legitimate sources of scientific insight or regulatory guidance. This book challenges that marginalization by demonstrating how ceremony, story-sharing, and land-based learning function as sophisticated systems of ecological observation and governance. These insights directly inform Chapter 7 and 8's recommendations for universities, environmental organizations, and research institutions to restructure knowledge production and policy engagement.

Third, relational ethics are largely absent from dominant climate governance frameworks. Climate policy continues to operate through extractive and instrumental logics that treat land, water, and energy as resources rather than living relations. Ethical considerations are often reduced to procedural consultation rather than reciprocal responsibility. This book advances a relational ethical framework rooted in Indigenous Traditional Land-Based Knowledge, reframing climate crisis as a rupture of relationships. This ethical orientation underpins the accountability-focused policy directions articulated in Chapter 6,7, 8, emphasizing long-term relational governance, community consent, and responsibility to future generations. Together, these contributions position the book as a critical intervention linking Indigenous theory, land-based practice, and actionable policy pathways for rethinking water and energy sustainability beyond colonial and technocratic paradigms.

CONTEXT: A POLYCRISIS OF LAND, WATER, AND COMMUNITY

This framework is deeply informed by the lived realities of Indigenous communities across Western and Northern Canada who are navigating a complex and interconnected polycrisis. Climate change is intensifying at a pace that threatens the stability of ecosystems, food systems, and cultural practices. At the same time, systemic colonial violence continues to undermine Indigenous governance, displace communities, and disrupt land-based knowledge transmission. These crises are not discrete—they are cumulative and mutually reinforcing.

Changing hydrological patterns, melting permafrost, rising water temperatures, shrinking wetlands, and prolonged droughts have transformed traditional land and water systems (Whyte, 2020; McGregor, 2021). Wildfires now burn longer, hotter, and across wider regions than in previous generations. These environmental changes limit access to cultural camps, disrupt wildlife migration routes, alter fishing cycles, and reduce the availability of traditional medicines and foods.

Alongside these climate disruptions, extractive infrastructures—hydroelectric dams, mining operations, oil sands developments, and pipelines—continue to reshape Indigenous territories. These infrastructures were overwhelmingly built without Indigenous consent, disrupting ecosystems, displacing communities, and contaminating water systems that have supported cultural and ceremonial life for generations (Baker et al., 2022; Latulippe & Klenk, 2020). The physical presence of pipelines, dams, and extraction sites symbolizes the political and legal structures that have historically denied Indigenous nations the power to govern their own waters.

Water insecurity remains a persistent crisis. More than thirty Indigenous communities in Canada continue to live under long-term boil water advisories (Indigenous Services Canada, 2024). Many more experience energy poverty, where access to affordable, clean, and culturally appropriate energy is limited by systemic inequities (Holcka et al., 2022). These are not technical failures they are the material outcomes of colonial systems that prioritize extraction over relationship, profit over community care, and national interests over Indigenous sovereignty.

CULTURAL AND CEREMONIAL CONSEQUENCES

Environmental injustice has profound cultural consequences. As traditional land-based practices become harder to sustain, language loss accelerates, Elders lose opportunities to pass on teachings, and youth experience disconnection from cultural identity. Ceremonial practices tied to specific water bodies—spring offerings, fasting camps, water songs, and seasonal gatherings—are interrupted when those water bodies are polluted or inaccessible (Borrows, 2020; Craft, 2021).

Ceremonial grief arises when communities witness the death of lakes, the drying of rivers, or the contamination of drinking water. These losses are not metaphorical—they represent spiritual ruptures that impact mental, emotional, and community health. Yet even within these painful realities, Indigenous communities continue to lead cultural and ecological renewal.

RESURGENCE AS METHOD, GOVERNANCE, AND ADAPTATION

Indigenous nations are not passive recipients of climate crisis. They are leaders in climate adaptation rooted in land-based governance. Across Western and Northern Canada, Elders, youth, and community leaders are organizing water walks, cultural camps, language revitalization programs, land-water monitoring, and community-led environmental initiatives. These actions regenerate relationships, revitalize ceremonies, and assert Indigenous sovereignty.

This book therefore calls for a paradigm shift: climate adaptation must not be confined to technical interventions or policy reforms. It must be rooted in cultural resurgence, relational justice, and Indigenous governance. Within Indigenous worldviews, water and energy are not resources—they are living relatives whose care demands ethical, ceremonial, and political responsibility.

VISION FOR THE FUTURE

This book is not merely a critique of past harms or a recounting of colonial disruptions—it is, above all, a ceremonial offering and a forward-looking vision for sustainable futures rooted in Indigenous life, law, and land-based relationality. At a time when climate change intensifies ecological, cultural, and political instability, this work affirms that true sustainability cannot emerge from technocratic innovation alone. Instead, it must arise from renewed relationships—among peoples, communities, nations, waters, lands, ancestors, and the more-than-human world. The future we imagine is grounded in the embodied and ancestral knowledges of Indigenous nations who have maintained relational governance systems for thousands of years despite ongoing colonial intrusions.

The vision offered here is relational rather than prescriptive. It is not a rigid blueprint or external plan—it is a living, breathing set of commitments grounded in Indigenous epistemologies. We envision a world where water and energy governance is guided first and foremost by Indigenous nations, whose laws and teachings articulate responsibilities long ignored by settler states. In this future, sustainability is measured not by economic efficiency or market growth, but by the health of waters, the flourishing of languages, the strength of kinship networks, and the vitality of future generations. This future is rooted in Indigenous worldviews that understand the Earth as a relative rather than a resource, and sustainability as a process of returning to right relationship.

We also envision a future where land-based education is recognized as essential to climate adaptation and environmental healing. In this future, Indigenous language immersion schools—such as *kâ-nêyâsihk mîkiwâhpa*—are not peripheral experiments, but fully supported institutions where cultural continuity, ecological ethics, and land-based responsibilities are taught holistically. Children grow up learning the land-based teachings of their ancestors: when to offer tobacco, how to listen to water, how to read the migration of geese, how to gather medicines respectfully, and how to maintain community responsibilities through ceremony. Land-based learning becomes the primary mode of preparing youth—not only for climate disruption, but for cultural resurgence, governance leadership, and environmental stewardship.

This book seeks to uplift those futures already coming into being. Across Indigenous territories, Cree, Dene, Anishinaabe, Métis, Inuit, and other nations are leading transformative work: solar microgrid systems that reduce dependence on colonial energy regimes; grandmother- and youth-led water walks that re-establish ceremonial relationships with rivers; community-based land camps that restore cultural teachings; and language revitalization movements that reconnect peoples to ancestral memory. These are not marginal initiatives. They are vivid expressions of Indigenous survivance, relational governance, and ecological intelligence. They represent pathways toward a world where sustainability is practiced as ceremony, responsibility, and kinship, not as resource extraction or crisis management.

To realize this future, however, requires profound shifts in institutional and political structures. Universities, government agencies, environmental organizations, and settlers must move beyond symbolic gestures of inclusion. They must take up their relational responsibilities by supporting—not appropriating—Indigenous knowledge systems. This means committing to long-term community-led partnerships grounded in reciprocity, transparency, and respect. It means ensuring that funding supports Indigenous-controlled initiatives rather than academic or governmental extraction. It means recognizing that climate justice is inseparable from treaty justice, and that meaningful reconciliation requires land return, shared governance, and culturally grounded decision-making.

Energy transition cannot succeed without cultural renewal. Environmental protection cannot be isolated from the restoration of Indigenous laws and languages. Climate adaptation must be rooted in community healing, ceremony, and the revitalization of traditional lifeways. These principles challenge dominant environmental governance frameworks, which often prioritize market-based incentives, technological innovation, or expert-driven planning. Indigenous visions for the future remind us that genuine sustainability cannot be achieved within paradigms that continue to marginalize Indigenous authority and perpetuate colonial relationships to land.

Our vision is not utopian. It is grounded in the daily practices, land-based experiences, and ceremonial teachings of Indigenous communities who continue to protect the sacred despite centuries of displacement, suppression, and environmental harm. These communities offer models of resilience and possibility not only for Indigenous peoples, but for all who share these lands and waters. Their teachings illuminate pathways toward relational governance that can address the climate crisis at its roots: not merely through adaptation or mitigation, but through restoring the ethical foundations of human-land relationships.

In sharing this book, we invite readers into this vision—not as observers, but as kin, participants, and relational partners. Readers are invited to walk with Indigenous teachings, to listen deeply to water, to learn from Elders, and to understand climate crisis not simply as a scientific problem but as a relational rupture requiring collective healing. Let this book be a doorway. Let it open toward futures shaped by ceremony, kinship, and responsibility. Let it accompany all those who carry water, tend sacred fires, harvest medicines, honour ancestors, and speak to the trees. As Elders often remind us, the future is not something we predict—it is something we grow through good relations, good intentions, and good paths.

REFERENCES

Arsenault, R., Diver, S., McGregor, D., Witham, J., & Bourassa, C. (2018). Contextualizing water justice: Indigenous water governance in settler colonial Canada. *Water (Basel)*, *10*(5), 1–19. PMID: 30079254

Baker, J., Smith, L., & Richmond, C. (2022). Colonial infrastructures and Indigenous water insecurity in Canada. *Global Environmental Change*, *72*, 102–118.

Bang, M., Marin, A., Faber, L., & Suzukovich, E. (2012). Reframing science education from Indigenous worldviews. *Cultural Studies of Science Education*, *7*(2), 535–547.

Borrows, J. (2020). *Law's Indigenous ethics*. University of Toronto Press.

Craft, A. (2021). *Anishinaabe Nibi inaakonigewin: The legal principles of water*. University of Manitoba Press.

Datta, R. (2015). A relational theoretical framework and meanings of land, water, and air: A decolonial, Indigenous approach to environmental justice. *Journal of Environmental Studies and Sciences*, *5*(3), 1–12.

Datta, R. (2018). Decolonizing both researcher and research and its effectiveness in Indigenous research. *Research Ethics*, *14*(2), 1–24. DOI: 10.1177/1747016117733296

Hoicka, C. E., Savic, K., & Campney, A. (2022). Indigenous energy sovereignty in Canada: Policy, power, and colonialism. *Energy Research & Social Science*, *89*, 102–112.

Houle, K. (2022). Indigenous ecological grief and climate pandemic realities. *Environmental Humanities*, *14*(1), 35–56.

Indigenous Services Canada. (2024). *Long-term drinking water advisories on public systems on reserves*. Government of Canada. https://www.sac-isc.gc.ca

Kimmerer, R. W. (2013). *Braiding sweetgrass: Indigenous wisdom, scientific knowledge, and the teachings of plants*. Milkweed Editions.

LaDuke, W. (2020). *To be a water protector: The rise of the Wiindigoo slayers*. Fernwood Publishing.

Latulippe, N., & Klenk, N. (2020). Making room and moving over: Knowledge co-production, Indigenous knowledge sovereignty and the politics of global environmental change decision-making. *Current Opinion in Environmental Sustainability*, *42*, 7–14. DOI: 10.1016/j.cosust.2019.10.010

Levac, L., McMurtry, J. J., & Walters, D. (2023). Energy poverty, equity, and Indigenous-led transitions. *Energy Policy*, *176*, 113–127.

Lewis, M., & Datta, R. (2023). Resurgence and relational governance in Indigenous climate leadership. *Sustainability Science*, *18*(4), 1123–1137.

McCarty, T., & Nicholas, S. (2014). Language loss in Indigenous communities and the environmental crises of meaning. *Annual Review of Anthropology*, *43*, 1–17.

McGregor, D. (2012). Traditional knowledge: Considerations for protecting water in Ontario. *International Indigenous Policy Journal*, *3*(3), 1–21. DOI: 10.18584/iipj.2012.3.3.11

McGregor, D. (2021). Indigenous environmental justice and sustainability. *The Canadian Journal of Native Studies*, *41*(1), 85–110.

Simpson, L. (2014). Land as pedagogy: Nishnaabeg intelligence and rebellious transformation. *Decolonization*, *3*(3), 1–25.

Tuck, E., & Yang, K. W. (2012). Decolonization is not a metaphor. *Decolonization*, *1*(1), 1–40.

Whyte, K. P. (2017). Indigenous climate change studies: Indigenizing futures, decolonizing the Anthropocene. *English Language Notes*, *55*(1–2), 153–162. DOI: 10.1215/00138282-55.1-2.153

Whyte, K. P. (2020). Too late for Indigenous climate justice? *Wiley Interdisciplinary Reviews: Climate Change*, *11*(1), 1–8. DOI: 10.1002/wcc.603

Wilson, S. (2008). *Research is ceremony: Indigenous research methods*. Fernwood Publishing.

Wilson, S. (2020). Learning from the land: Indigenous paradigms and research relationships. *Journal of Indigenous Studies*, *15*(2), 45–60.

Chapter 2
Indigenous Traditional Water Knowledge and Practice as Environmental Sustainability

ABSTRACT

Chapter Two reframes water crisis as a profound cultural, spiritual, and relational rupture for Indigenous communities in Western Canada. Rather than viewing water scarcity, contamination, and hydrological instability as isolated environmental problems, the chapter demonstrates how water crisis accelerates the erosion of Indigenous languages, knowledge systems, ceremonial practices, and intergenerational relationships. Drawing from Elders' testimonies, land-based narratives, and community experiences, the chapter illustrates six thematic areas through which water crisis disrupts Indigenous lifeways: the destabilization of land-based knowledge systems, economic displacement, language loss, ecological crisis, cultural fragmentation, and threats to community well-being. Yet, the chapter also highlights Indigenous-led responses—such as land-based education, language immersion, and youth leadership—that revitalize relational ethics and restore responsibility to water. Thus, the chapter positions Indigenous Traditional Water Knowledge as essential to environmental sustainability.

WATER CRISIS AS THE EROSION OF INDIGENOUS LIFEWAYS

Water is not merely a natural resource; it is life and a sacred presence within many Indigenous worldviews to the communities that we had opportunity to learn. Across Turtle Island, Indigenous nations have long held critical relationships with water, governed not by extraction but by responsibility, reciprocity, and ceremony. Water connects generations, sustains oral traditions, nourishes bodies and ecosystems, and carries teachings through rivers, lakes, snow, and rain. Yet, water crisis and settler-colonial economic structures increasingly threaten these relationships. We examined how traditional water knowledge and practice function as pillars of environmental sustainability, and how these practices are endangered—and revitalized—in the context of environmental disruption, cultural loss, and economic displacement (Arsenault et al. 2018).

As we learned that water crisis poses an existential threat to Indigenous water relationships and, by extension, to Indigenous ways of life. The rapid transformation of hydrological cycles—flooding, droughts, contaminated aquifers, and receding glaciers—destabilizes the very conditions necessary for Indigenous communities to engage in water-based ceremonies, knowledge transmission, and ecological governance. For many Indigenous cultures, water is not only a site of sustenance but a teacher, healer, and carrier of language. It is embedded in the structure of stories, placenames, songs, and kinship systems.

When rivers run dry or lakes freeze irregularly, Elders can no longer teach seasonal fishing, canoe carving, or the timing of water ceremonies. A knowledge keeper from North Cree First Nation explains: "We used to gather on the lake every spring, and that's when the language came alive. We taught the kids, we told stories, we harvested fish. Now, the water comes late, the ice stays longer, and the gathering doesn't happen." This shift reflects a deeper dislocation: water crisis interrupts the seasonal rhythms that hold Indigenous knowledge and language in place. Without stable and predictable water patterns, the intergenerational transfer of land- and water-based knowledge is fractured.

LANGUAGE LOSS AND WATER DISRUPTION

The erosion of traditional water systems is deeply connected to the erosion of Indigenous languages. Language and water are relational systems: both are fluid, cyclical, and rooted in specific geographies. Many Indigenous languages contain highly specialized vocabularies for describing the nuances of water—its flow, taste, temperature, spiritual significance, and ecological behavior. These words often lack direct translations in English or French, and their usage is contingent on direct engagement with water systems. As one Elder remarked, "In the last 10 to 15 years, we've lost many water words. Words for different kinds of ice, words for the way the river turns in spring, words for the silence after the rain. If you don't live with the water, you don't speak those words anymore." This testimony illustrates that water crisis, by altering or degrading water environments, is directly responsible for accelerating language loss. Language cannot exist in abstraction; it must be spoken in context, and that context is being reshaped or erased by ecological instability. Furthermore, the disappearance of ceremonial water practices—such as water walks, spring blessings, and snow-harvest teachings—leads to linguistic impoverishment. Ceremonies are often conducted in the Indigenous language and are occasions for youth to hear, learn, and practice ancestral words. As these practices become less frequent or are relocated away from water due to environmental hazards, the linguistic dimension of culture suffers. Thus, water degradation becomes both a physical and a cultural loss.

Recent demographic data underscore the scale and intergenerational nature of Indigenous language loss occurring alongside environmental disruption. According to Statistics Canada census data, fewer than 16% of Indigenous peoples in Canada report conversational ability in an Indigenous language, with fluency heavily concentrated among Elders aged 65 and older, while rates drop sharply among adults and youth (Statistics Canada, 2021). In many First Nations communities, fewer than 10% of children and youth are fluent speakers, despite the presence of strong cultural knowledge among older generations. Scholars have demonstrated that language transmission is most resilient where daily land-based practices—such as fishing, gathering, and water-centered ceremonies—remain intact (McIvor, 2020; McCarty & Nicholas, 2014). As water systems become contaminated, unpredictable, or inaccessible, the ecological contexts in which Indigenous languages are spoken are disrupted, accelerating intergenerational language loss. These trends illustrate that language decline is not only a legacy of residential schooling and colonial policy, but also a contemporary outcome of environmental degradation that fractures land-language relationships.

Water crisis is not only a crisis of rising global temperatures, melting glaciers, or shifting weather systems. For many Indigenous communities, particularly those whose identities and livelihoods are intricately connected to land-based knowledge systems in Western Canada, water crisis is an existential threat to cultural survival, linguistic continuity, and collective well-being. This chapter positions water crisis as a vehicle accelerating the destruction of Indigenous land-based cultural frameworks, but also as a settler colonial force that exposes the continuing strength of Indigenous knowledge, language, and relational practices (Schuster et al. 2019).

The goal of this chapter is to document how colonial infrastructures, extractive industries, and environmental degradation have disrupted Indigenous peoples' relationships with water, resulting in not only ecological harm but also cultural, emotional, and spiritual trauma. Its objectives include highlighting water insecurity as a relational rupture, gathering testimony from Elders and community members about the cultural significance of water, and illustrating how land-based pedagogies, ceremonies, and youth-led actions restore responsibility and relational ethics.

Reframing water governance as spiritual and cultural practice, the chapter reveals water as both a teacher and a site of resurgence.

Indigenous peoples around the world maintain strong relationships with their lands, waters, and non-humans. These relationships are interconnected with Indigenous languages, embedded in land-based ceremonies, enacted through traditional land-based knowledge, and transferred through intergenerational traditional storytelling (Datta, 2018; Simpson, 2014). Yet these sustainable infrastructures are being destabilized by climate-induced disruptions: extreme weather, droughts, floods, forest fire, animal migration shifts, and ecosystem collapse are not merely environmental phenomena—they are cultural traumas (Whyte, 2017; Simpson, 2014). As climate risks combinations the impacts of settler colonialism, it accelerates the loss of Indigenous languages, alters traditional food systems, interrupts ceremonial practices, and weakens intergenerational relationships. These shifts pose extreme threats to Indigenous identity, sovereignty, and survivance. This chapter reframes water crisis from the standpoint of Indigenous peoples, particularly drawing on stories from communities in Western Canada. Rather than treating environmental degradation as a purely external threat, this chapter situates it within the historical and ongoing processes of settler colonialism, resource extraction, and settler-capitalist expansion. These forces have long disrupted Indigenous ways of living and knowing, and water crisis, as both a symptom and accelerant of these processes, is deepening structural inequalities and cultural dislocation.

Figure 1. Individual and collective elders and knowledge-keepers water stories

At the center of this chapter is an inquiry into how the water crisis intersects with Indigenous sustainable water knowledge and practice in six key areas: the settler colonial impact on traditional knowledge systems; the economic displacement caused by environmental and colonial pressures; the loss of Indigenous languages tied to land and ecology; the emotional and spiritual toll of witnessing ecosystem collapse; the role of Indigenous languages in articulating ecological ethics; and the resurgence of Indigenous-led educational practices that respond to both climate and cultural crises. This framing draws from the lived experience, knowledge, and stories of Indigenous Elders, language speakers, educators, and youth who have articulated how water crisis is reshaping their relationships with land, language, and community. Their voices—recorded through traditional story sharing, land-based ceremonies, and land-based practices—are centered throughout the chapter to challenge western forms of environmental research. Rather than speak about Indigenous communities, this chapter amplifies how Indigenous peoples themselves understand, experience, and respond to environmental transformation.

Importantly, the chapter does not position Indigenous communities solely as victims of water crisis. Rather, it emphasizes their role as knowledge holders, cultural protectors, and leaders in climate adaptation. Indigenous responses—such as language engagement programs, community land-based education, cultural camps, and land stewardship practices—are examples of both resilience and resistance. These practices do not simply aim to "adapt" to water crisis in a technocratic sense but instead seek to revitalize and restore Indigenous relationships with the land, based on principles of reciprocity, kinship, and accountability (Kimmerer, 2013). Therefore, the objectives of this chapter are threefold:

1. To examine how the water crisis disrupts Indigenous water practices and traditional water knowledge systems. Through the destabilization of seasonal cycles, species migration, and environmental predictability, water crisis undermines the contexts necessary for intergenerational learning and ceremonial practice.
2. To explore the relationship between environmental degradation and economic insecurity, particularly in relation to water sovereignty, employment, and migration. This section highlights how land-based economies are threatened by both ecological decline and colonial economic policies, leading to cultural and spatial dislocation.
3. To highlight Indigenous-led responses that center land-based water sustainability education, language revitalization, and healing practices as strategies for cultural survival. In the face of cultural and ecological crisis, Indigenous communities are actively developing frameworks for water justice that are interconnected in traditional ecological knowledge, relational ethics, and intergenerational responsibility.

These objectives are explored through six interconnected themes and seventeen subthemes, which draw from decolonial story sharing, lived narratives, and participatory research collaborations with Indigenous communities. Each theme maps how water crisis is experienced not as an abstract or future threat, but as a lived reality disrupting the daily rhythms of language use, cultural practice, economic life, and spiritual connection. The chapter begins by examining how water crisis directly threatens Indigenous languages and knowledge systems, both of which are rooted in land-based observation and seasonal patterns. The destabilization of these patterns—through unpredictable weather and species decline—interrupts the contexts necessary for language use and oral storytelling. As language fades, so too does the cultural memory and epistemic diversity it holds (McCarty & Nicholas, 2014). Indigenous languages are not simply tools for communication; they are ontologies that reflect how people understand and relate to the world. Losing language means losing a worldview. Next, the chapter explores how colonial economic structures exacerbate the impacts of water crisis, particularly through unemployment and forced migration. Many Indigenous communities face systemic barriers to employment on reserves, where opportunities are limited and economic development is restricted by policies such as the Indian Act. As a result, people—especially youth—migrate to urban centers, often at the cost of cultural and linguistic disconnection. This migration is not voluntary; it is an effect of both economic exclusion and environmental unsustainability. The result is a further fragmentation of community life, traditional labor, and cultural mentorship. Subsequent sections examine how environmental degradation impacts emotional well-being, family cohesion, and cultural continuity. Testimonies throughout the chapter speak to the anxiety and spiritual exhaustion associated with water crisis. Elders report fewer opportunities to pass on teachings; youth describe the disappearance of animals they once knew the names for; parents lament the loss of seasonal food and medicines that once structured family routines. This emotional toll is a form of climate crisis—what some scholars refer to as "solastalgia"—but in Indigenous contexts, it is also a political and spiritual crisis.

Yet, alongside these stories of disruption are stories of resurgence. Indigenous communities are creating land-based educational initiatives, developing immersive language programs, and revitalizing ceremonial life in the face of adversity. These programs—often community-led and Elder-guided—offer powerful counter-narratives to dominant climate solutions that focus solely on technology and mitigation. For Indigenous peoples, the land is not merely a site of extraction or carbon sequestration; it is a teacher, a relative, and a source of ethical life. Climate adaptation, therefore, must begin with cultural regeneration. Reframing water crisis as a cultural and spiritual emergency, this chapter contributes to a growing body of Indigenous scholarship that calls for an epistemic shift in how we understand and address the water crisis (Wildcat, 2009; Tuck, McKenzie, & McCoy, 2014).

Indigenous peoples are not just disproportionately impacted by water crisis—they also hold knowledge systems and relational practices that offer critical pathways forward. These pathways include not only scientific observations but also stories, songs, ceremonies, and teachings that have sustained communities through generations of upheaval. Therefore, this chapter argues that addressing water crisis in Indigenous contexts requires more than technical adaptation strategies—it demands cultural justice, linguistic revitalization, and spiritual healing. It requires recognizing Indigenous sovereignty, supporting community-led education, and restoring relationships to land that have been severed by colonization and capitalist extraction. Climate justice is cultural justice (Ford et al. 2020). And Indigenous resurgence is central to the collective survival of both ecosystems and cultures.

DECLINE OF TRADITIONAL WATER KNOWLEDGE

Traditional water knowledge is holistic, passed orally from Elders to youth through direct experience with rivers, lakes, rain, and snow. It includes ways of predicting seasonal floods, understanding fish migration, interpreting water clarity, and preparing for drought. This knowledge is not only ecological—it is spiritual, ethical, and embodied. Yet, with water crisis disrupting hydrological patterns, the reliability of this knowledge is under threat.

One young community member from Treaty 6 territory laments: "We don't study the water anymore. That was the job of the Elders. But many of them have passed on, and the water isn't the same. It's harder to know when to fish, where to gather water, or how to read the sky." The quote highlights a generational rupture in the continuity of water knowledge. The knowledge was not written down; it lived in people, places, and patterns. Without those patterns—due to water crisis—the knowledge becomes fragile. The decline of water-based knowledge systems also impacts community safety and resilience. In the past, Elders could read the land and water to warn of flooding or drought; today, those cues are no longer legible. Western science may provide satellite predictions, but it does not replace the depth and specificity of local water knowledge accumulated over generations.

COLONIAL ECONOMIES, UNEMPLOYMENT, AND DISCONNECTION FROM WATER

The intersection of colonial economic systems and water crisis exacerbates the loss of Indigenous water practices by forcing people off the land and into urbanized labor markets that do not value traditional knowledge. In many reserves, economic opportunities are limited to teaching, administration, or health care. As one community member explains, "There is no water-based economy anymore. Fishing is not viable. Waterways are polluted. Youth have to leave to find work. And when they leave, they lose that connection to the water."

Unemployment in Indigenous communities is not simply a lack of jobs—it is a lack of land-based, water-centered livelihoods that sustain culture. The colonial imposition of capitalist employment structures replaced Indigenous economies of reciprocity, subsistence, and seasonal labor. Where once people fished, gathered wild rice, or conducted water ceremonies as integral parts of economic and spiritual life, they are now relegated to survival on welfare or migration to urban centers.

This migration has significant cultural consequences. When people move away from their water bodies, they lose access to the spiritual, linguistic, and educational dimensions of water. A young man from a Northern community shared: "When I moved to the city, I stopped speaking the language. There was no lake. No one to fish with. No one to say those words with." His experience illustrates the cascading effect of unemployment, migration, and disconnection from place-based knowledge.

Water, Technology, and Cultural Displacement

Technological advancement has further widened the gap between youth and traditional water practices. With the rise of digital media, many youths spend less time on the land or water. As one Elder noted, "Before, kids were always down at the river—learning, playing, listening. Now, they're indoors on their phones. They don't know the water like we did." This disconnection is not a simple matter of lifestyle change; it reflects a broader cultural disorientation where water, once central to daily life, becomes distant and abstract.

Technology, in the absence of cultural anchoring, displaces the knowledge systems that connect youth to water. It is not that technology is inherently harmful, but that it often reproduces colonial values of control, consumption, and convenience—values that stand in stark contrast to Indigenous teachings about water as sacred and alive. To reverse this disconnection, communities are beginning to reintroduce water walks, canoe camps, and digital storytelling rooted in water-based teachings.

Water Language Keepers and the Need for Recognition

Fluent speakers who hold water-based knowledge are invaluable cultural leaders. They carry words for fish, river bends, ice patterns, and water songs. Yet, in current economic systems, these speakers are often under-recognized and under-compensated. A community educator explains: "Fluent speakers have so much to offer, especially about the water. But they aren't given teaching roles. If we want to keep the language and the water knowledge, we have to support them."

Supporting water language keepers requires structural change: creating roles for them in schools, language nests, water monitoring programs, and environmental policy. Their knowledge is not only cultural but practical—it can guide water stewardship, inform land use decisions, and help communities adapt to water crisis. Water language is not a relic; it is a living knowledge for environmental governance.

INDIGENOUS LAND-BASED LANGUAGE THREATENED BY WATER CRISIS

Indigenous languages are more than tools for communication; they are repositories of ecological wisdom, cultural memory, and spiritual relationships with land and water. These languages are inherently rooted in place, drawing meaning from specific ecosystems, seasonal cycles, species behavior, and ceremonial practices. As water crisis accelerates environmental decline, it disrupts the very landscapes and relationships that sustain Indigenous languages. This erosion is not merely linguistic—it is cultural, spiritual, and existential. Water crisis destabilizes the conditions necessary for language to thrive. It transforms the land in ways that interrupt long-standing ecological rhythms, from animal migrations and forest health to weather patterns and food cycles. These environmental changes are deeply intertwined with cultural knowledge systems, many of which are encoded in language. As plants and animals disappear, as seasons become unpredictable, and as traditional gathering and hunting practices are curtailed, so too vanish the words, metaphors, and teachings tied to these elements. In this way, water crisis is accelerating the erosion of Indigenous language by severing its connections to lived land-based experience.

Language as an Indicator of Cultural Health

One of the most illuminating insights shared by Indigenous knowledge keepers is the recognition that language fluency reflects the overall health of a culture. Language is an ontological system—it is a way of being in the world. When language fades, it signals a weakening of spiritual practices, intergenerational bonds, and relationship with the land. As one Elder explained, "Language is a signifier of the well-being of a culture. If you have children who speak the language, that's a healthy culture of people."

This observation reminds us that language is not simply a matter of vocabulary—it is the foundation of identity, memory, and collective belonging. The loss of Indigenous languages, therefore, is not an isolated consequence of globalization or assimilation policies; it is also deeply linked to environmental degradation. As language disappears, so does a people's capacity to express their history, their connection to land and water, and their worldview. The health of the language reflects the health of the people, the land, and the cultural systems that sustain them.

Animal Migration as an Early Warning Sign

Indigenous languages are shaped by intricate relationships with non-human relatives. Animals, in particular, hold central places in stories, teachings, seasonal cycles, and ceremonial protocols. The migration, behavior, and spiritual significance of animals provide both vocabulary and cultural narrative. When animals begin to disappear or shift their migration routes due to water crisis, they take with them the words and knowledge systems associated with their presence.

As one community member expressed, "How do we know the land is struggling? We know by the migrations that are happening." This perspective illustrates how animal migration is not only an ecological event but also a linguistic and cultural indicator. When a particular bird no longer returns in spring, the songs, teachings, and words connected to its presence may fall into disuse. Language fragmentation follows ecological fragmentation. Thus, the erosion of biodiversity leads directly to the erosion of linguistic and cultural diversity.

Limited Time for Cultural Communication

Another cascading impact of water crisis on Indigenous languages is the reduction in time available for land-based cultural communication. Extreme weather events, heatwaves, and unpredictable seasonal patterns increasingly prevent families and communities from gathering outdoors, engaging in ceremonies, or spending extended time on the land. These disruptions limit opportunities for intergenerational language transmission. One community member poignantly shared, "When the heat wave hits, we have less time outside, less time in the water, less time just to be together. So it's like everyone is cooped up in their own spaces." This statement reveals how climate adaptation is not neutral—it restructures daily life in ways that diminish communal interaction, storytelling, and cultural expression. Language thrives in the space between people and place. When those spaces are compromised, language transmission is curtailed. Cultural knowledge requires time, presence, and rhythm—all of which are being shortened by ecological crisis.

Loss of Traditional Forest and Ecological Reference Points

The disappearance of specific plants, animals, and landforms due to environmental decline also contributes to the hollowing out of Indigenous language. Traditional ecological knowledge includes highly specialized vocabularies for describing medicinal plants, forest patterns, animal behaviors, and seasonal indicators. These words often lack equivalents in colonial languages and are inseparable from the land-based practices they describe. An Elder reflected on this loss with sorrow: "I don't even remember what it was called in our language. That's a good example of how language disappears—if you're not speaking it, you're not going to remember it." This quote captures the pain of witnessing not just ecological loss, but also the silencing of language that once described and honored those beings. When nature disappears, language becomes orphaned. Words become abstract, disconnected from lived experience, and eventually forgotten. The result is not only linguistic erosion but cultural grief.

Food Insecurity and Cultural Transmission

Food systems are deeply embedded in Indigenous cultural practices. Gathering, hunting, preparing, and sharing food involve specific language, ritual, and relational protocols. These activities transmit not only vocabulary but also values such as respect, reciprocity, and gratitude. When water crisis disrupts food systems—through drought, declining animal populations, or changing growing seasons—it also disrupts the transmission of language and cultural teachings. As one speaker noted, "Gathering around the table as a family is a symbol of health. It keeps the kids and the family together." This gathering is not only about nutrition—it is about ceremony, language, and cultural cohesion. When traditional foods become scarce or inaccessible, families lose a central site of cultural expression. Vocabulary associated with food preparation, seasonal knowledge, and social structure falls into disuse. As with other dimensions of land-based knowledge, the survival of language depends on the survival of the practices in which that language is embedded.

Ecological Change and Knowledge Disruption

Water crisis also brings with it intense emotional and spiritual consequences. For Indigenous communities, the land is not just physical territory—it is kin, teacher, and healer. The disruption of ecosystems causes anxiety and disorientation. It also interrupts the transmission of ecological knowledge, much of which is conveyed through experiential learning and embodied storytelling.

One participant described how environmental change affects mental well-being: "The seasons are changing, and I notice my own behavior changes. It affects my mental health." This quote illustrates that land-based stress is not only ecological; it is deeply somatic and spiritual. Emotional fatigue, in turn, diminishes the energy and presence needed to engage in cultural teaching. Without environmental continuity, traditional teachings become speculative, and the embodied aspects of knowledge transmission are lost. In this way, ecological instability undermines cultural stability.

Language as an Ecological System

What emerges from all these subthemes is the recognition that Indigenous languages are ecological systems. They are interdependent, contextual, and rooted in specific relationships with land, water, and more-than-human beings. When these ecosystems are disrupted, the languages that articulate and sustain them also suffer. Water crisis does not only disrupt weather or displace species—it silences stories, fragments intergenerational memory, and collapses the scaffolding of cultural life.

Protecting Indigenous languages, therefore, cannot be separated from protecting the land. Efforts to revitalize language must include support for land-based practices, ceremonial life, and cultural education rooted in environmental stewardship. Indigenous language revitalization is not only a cultural act—it is an ecological intervention.

Indigenous languages are not simply systems of communication; they are ecological philosophies, spiritual guides, and relational teachings embedded in words, metaphors, and place-based expressions. These languages hold within them generations of observations, ethical instructions, and ceremonial responsibilities tied to the land, water, animals, and climate. They teach not only how to name the world, but how to live in right relationship with it.

In the context of water crisis, Indigenous languages offer a vital framework for responding to environmental crisis—not just through adaptation, but through transformation of worldview. These languages contain ecological logic and spiritual protocols for sustainability. However, as water crisis accelerates the erosion of ecosystems, it also destabilizes the cultural and environmental contexts that give Indigenous languages their meaning, relevance, and vitality.

Language Loss as Disconnection from Ecological Ethics

When Indigenous languages fade, the impact goes beyond the loss of words—it severs ties to the Earth's rhythms, ethics, and responsibilities. Language is not only a cognitive tool, but a spiritual and cultural map for how to live with land and water. This is particularly evident in the subtheme concerning the loss of language as a driver of ecological disconnection. A knowledge keeper shared, "The ceremonies that we have require the language. It requires the language as the chants are all in the language. The prayers are all in the language." This statement illustrates that without the language; ceremonial life cannot be fully enacted. When chants, prayers, and oral teachings are lost, so too are the relational instructions embedded within them—how to harvest respectfully, how to give offerings, how to read the sky and understand the river's story. The loss of language means the loss of ethical direction. Disconnection from language becomes disconnection from the land's laws and sacred teachings.

Indigenous languages are inherently contextual. They are shaped by seasonality, geographic specificity, and active practice. Many words exist only in the context of particular ceremonies, ecological behaviors, or environmental patterns. When those contexts are altered by water crisis—when animals no longer return, plants stop growing in their traditional locations, or weather patterns shift—those words become unspeakable, unrecognizable, or obsolete. A community representative observed, "We've come away from land-based teachings so much that there are certain words that we just don't say anymore. Because they're only said in certain contexts." This insight reveals the fragility of context-bound language and the way water crisis disrupts not only ecosystems but also the vocabulary embedded within them. When snow does not fall when it used to, the words for the different types of snow—their textures, their sounds, their spiritual meanings—begin to disappear. Climate disruption erodes the linguistic scaffolding of cultural instruction. This phenomenon demonstrates that language is not just a tool for remembering the past—it is a means for behaving in the present. Indigenous languages encode knowledge about interdependence, respect, and ecological balance. As environmental change alters or eliminates the settings that give meaning to these teachings, the ability to live in accordance with ancestral ethics is increasingly compromised. In this sense, language loss is not only a cultural loss; it is also an ecological and moral one.

Indigenous Educational Responses to Language and Water crisis

Despite these immense challenges, Indigenous communities are leading innovative responses that confront the dual crisis of water crisis and language erosion. Central to these responses is the revitalization of land-based education. These educational practices do not separate language from land, or culture from climate—they teach them together, because they have always been interwoven. Land-based language immersion programs are one of the most powerful strategies communities are using to restore language vitality. These programs involve elders, fluent speakers, and youth participating in cultural activities—such as fishing, canoeing, harvesting, and storytelling—in the language, on the land. These practices reconnect words with their environmental referents, while reinforcing relational ethics and community knowledge.

An Indigenous educator reflected, "In the classroom, because I'm an educator, I am using the land-based language immersion method. And it is based on implements that are used outside in different contexts through land-based activities." This quote speaks to the necessity of reuniting language with land in educational spaces. When young people learn language through lived experience—touching snow, listening to the wind, carving a paddle—they do not just learn words; they learn responsibilities, relationships, and ways of being that align with Indigenous epistemologies. Community-led language education is not merely about preservation—it is about resurgence. It is about restoring the foundations of cultural and ecological knowledge so that Indigenous futures can be imagined and enacted with strength and clarity. These programs make language vibrant and meaningful again, not in a museum, but in the everyday practices of life.

Language as Healing and Resistance

The recovery of Indigenous language through land-based learning is a powerful act of resistance and healing. It challenges the settler-colonial systems that have long devalued Indigenous knowledge, extracted Indigenous lands, and displaced Indigenous families from their territories. By reclaiming language, communities also reclaim stories, histories, and futures. Ultimately, this theme teaches us that Indigenous language is not simply about identity—it is about ecological survival. These languages carry the teachings needed to live in balance with a changing world. As water crisis threatens the land, water, and animals, it also threatens the languages that teach how to live in relationship with them. But in the face of loss, Indigenous educators, knowledge keepers, and youth are finding ways to restore the bonds between language, land, and life. In this way, education becomes ceremony. Language becomes medicine. And speaking in the language of the land becomes an act of renewal—not only of culture, but of Earth itself.

LEARNING REFLECTIONS

This chapter takes us into the center of the crisis where water crisis is not merely about environmental degradation, but about the deep, on-going settler colonization of Indigenous culture, identity, language, memory, and relationships to land and water. It shows that for Indigenous Peoples, water crisis is not an isolated ecological phenomenon; rather, it is an extension of historical colonial violence that continues to dismantle the conditions necessary for Indigenous lifeways to thrive. Through six interconnected themes and seventeen subthemes, this chapter offers a reflective exploration of how water crisis disrupts the relational fabric of Indigenous worlds—but also how Indigenous communities continue to assert their knowledge systems, languages, and resilience in the face of disruption.

One of the most powerful insights that emerged was the deeply relational nature of Indigenous language. Language is not simply a vehicle for communication—it is a living system interwoven with land, water, animals, and seasons. The connection between climate disruption and language loss becomes painfully evident. As weather patterns shift and become unpredictable, and as animal migration is disturbed or halted altogether, the words and concepts tied to those rhythms begin to vanish. As an Elder explains this reality with striking clarity: "When the animals leave, the words go with them." This speaks to a broader truth that language lives in context—it requires place, rhythm, and continuity. Water crisis unravels those contexts, and in doing so, weakens the scaffolding upon which intergenerational memory and cultural continuity rest.

Just as language is eroded, so too is the cultural labor and traditional economy that sustains community cohesion. Theme Two points to the profound implications of economic hardship within Indigenous communities. Unemployment is not just a material concern—it is a cultural crisis. As traditional work declines due to environmental degradation and colonial economic marginalization, youth are increasingly forced to migrate to urban centers in search of employment. This migration severs ties to the land, to Elders, and to land-based knowledge systems. Meanwhile, fluent speakers and cultural knowledge holders who remain in the community often lack support and formal recognition for their invaluable contributions. There are few, if any, paid opportunities for those who carry the language and teachings to pass them on. This highlights the structural failure of colonial economies to value Indigenous ways of knowing and being.

Various themes introduce us to another critical dimension of this water crisis. This is a pain not simply rooted in witnessing environmental change, but in feeling the emotional, physical, and spiritual toll of relationship loss. The loss of ecosystems is not just the loss of resources—it is the loss of ceremonies, memories, and teachings embedded in those places. People speak of exhaustion, depression, and a sense of spiritual dislocation as they watch once-familiar landscapes become unrecognizable. One participant candidly shares how their mental well-being is intimately tied to seasonal cycles and that when these rhythms are disturbed, their sense of self becomes unstable. Western scientific frameworks often ignore these emotional and spiritual dimensions of environmental loss. Yet, in Indigenous contexts, these are central to understanding what is truly at stake.

Despite these gaps, the chapter does not leave us in despair. On the contrary, various land-based adaption themes reminds us that Indigenous communities are actively responding through education, ceremony, and community-led innovation. Land-based learning and language immersion programs are not just acts of cultural preservation—they are strategies of decolonization and cultural resurgence. In classrooms, on traplines, at water ceremonies, and during seasonal camps, Elders and educators are teaching youth the language, ethics, and stories that emerge from the land. One educator describes how using land-based language immersion methods in her teaching reconnects students not only to their language but to the values and responsibilities embedded within it. Education, in this sense, becomes an act of reclamation and resistance. It is not remedial; it is revolutionary.

One of the central takeaways from this chapter is that climate justice must be understood as inseparable from cultural justice. The loss of language, the breakdown of intergenerational teaching, and the weakening of ceremonial life are not collateral damages—they are central dimensions of what water crisis looks like in Indigenous communities. If adaptation strategies are to be meaningful, they must include efforts to revitalize Indigenous languages, support traditional economies, and ensure the continuance of land-based education. Language revitalization is not a separate task from climate adaptation—it is part of the same work.

Indigenous land-based language in water introduces a crucial idea: that Indigenous languages are not simply descriptive, but instructive. They encode the ecological ethics, spiritual principles, and relational protocols necessary for sustainable living. When we lose Indigenous languages, we also lose a blueprint for how to live with the land rather than on it. This theme emphasized how ceremonies, chants, and teachings only make sense when spoken in the language of the land. Water crisis, by disrupting the environmental conditions required for these practices, threatens to render language—and the values it carries—obsolete. In this way, water crisis is not just eroding the physical world; it is also disorienting the spiritual and moral compass that guides human behavior in relation to the Earth.

This chapter also highlights the role of community-led education in healing and re-grounding cultural knowledge. Indigenous land-based education does more than transmit information—it rebuilds relationships. Language is taught alongside snow, story is told while harvesting medicine, ethics are learned in the act of preparing fish or building a canoe. These embodied forms of education are precisely what water crisis threatens—and yet, they are also the most promising instruments for cultural survival. Education becomes a place where misery is acknowledged, where memory is restored, and where youth are invited into ancestral responsibility. Throughout all themes, we are reminded that Indigenous peoples are not passive victims of water crisis. They are knowledge keepers, teachers, and innovators. They are actively engaging in the work of remembrance, repair, and renewal. They speak to the lake, even when the water is low. They teach the stories, even when the animals no longer return. They chant the songs, even when the forests are silent. This is not just resilience—it is a profound form of spiritual endurance and ethical clarity.

In conclusion, this chapter makes it clear that water crisis is not only about melting ice or rising temperatures. For Indigenous communities, it is about disappearing languages, disrupted teachings, broken relationships, and the challenge of remembering in a time of forgetting. But it is also about resurgence. It is about those who teach from memory, those who listen to the land even when it hurts, and those who rebuild futures with the means of ceremony, language, and love. To support Indigenous climate resilience is to support language teachers, canoe builders, medicine harvesters, storytellers, and ceremonial leaders. It is to understand that employment must include cultural traditional land-based practice that adaptation must include ancestral instruction, and that climate policy must include spiritual restoration. The lesson from this chapter is not one of despair but one of deep respect. Respect for those who carry the stories, who continue to teach the language of the land, and who refuse to let the songs be forgotten.

This chapter demonstrates that water crisis in Indigenous communities cannot be understood as an isolated environmental problem; it is a relational, cultural, and governance crisis that unfolds through language loss, economic displacement, ecological grief, and disrupted intergenerational teaching. As water systems change, the seasonal rhythms that sustain Indigenous languages, ceremonies, and land-based knowledge are destabilized, weakening the ethical frameworks that have long guided sustainable relationships with water. Yet, the chapter also makes clear that Indigenous communities are not passive victims of these transformations. Through land-based education, language immersion, ceremonial resurgence, and youth-led initiatives, communities are actively restoring responsibility to water and reaffirming Indigenous governance grounded in relational accountability. These practices reveal that cultural resurgence is not separate from environmental sustainability, it is its foundation. By situating water crisis within lived experiences of language, land, and community, this chapter prepares the ground for Chapter 3's deeper examination of ceremony as governance and water as a site of Indigenous law and leadership.

REFERENCES

Arsenault, R., Diver, S., McGregor, D., Witham, J., & Bourassa, C. (2018). Contextualizing water justice: Indigenous water governance in settler colonial Canada. *Water (Basel)*, *10*(5), 1–19. PMID: 30079254

Datta, R. (2018). Decolonizing both researcher and research: A critical, land-based approach to Indigenous research. *Research Ethics*, *14*(2), 1–24. DOI: 10.1177/1747016117733296

Ford, J. D., King, N., Galappaththi, E. K., Pearce, T., McDowell, G., & Harper, S. L. (2020). The resilience of indigenous peoples to environmental change. *One Earth*, *2*(6), 532–543. DOI: 10.1016/j.oneear.2020.05.014

Kimmerer, R. W. (2013). *Braiding sweetgrass: Indigenous wisdom, scientific knowledge, and the teachings of plants*. Milkweed Editions.

McCarty, T. L., & Nicholas, S. E. (2014). Language education, Indigenous revitalization, and building resilient communities. *Annual Review of Applied Linguistics*, *34*, 21–41.

Schuster, R., Germain, R. R., Bennett, J. R., Reo, N. J., & Arcese, P. (2019). Vertebrate biodiversity on indigenous-managed lands in Australia, Brazil, and Canada equals that in protected areas. *Environmental Science & Policy*, *101*, 1–6. DOI: 10.1016/j.envsci.2019.07.002

Simpson, L. B. (2014). Land as pedagogy: Nishnaabeg intelligence and rebellious transformation. *Decolonization*, *3*(3), 1–25.

Tuck, E., McKenzie, M., & McCoy, K. (2014). Land education: Rethinking pedagogies of place from Indigenous, postcolonial, and decolonizing perspectives. *Environmental Education Research*, *20*(1), 1–23. DOI: 10.1080/13504622.2013.877708

Whyte, K. P. (2017). Indigenous climate change studies: Indigenizing futures, decolonizing the Anthropocene. *English Language Notes*, *55*(1–2), 153–162. DOI: 10.1215/00138282-55.1-2.153

Wildcat, D. R. (2009). *Red alert! Saving the planet with Indigenous knowledge*. Fulcrum Publishing.

Chapter 3
Indigenous Water as Life

ABSTRACT

Chapter Three argues that Indigenous water relationships are grounded in ceremonial, spiritual, and ecological responsibilities that position water not as a resource but as a sacred, sentient relative. Through thirteen themes and nineteen subthemes, the chapter demonstrates how water ceremonies function as systems of governance, healing, and intergenerational cultural continuity. Indigenous communities' articulate water as life, emphasize the gendered responsibilities of grandmothers and ceremonial leaders, and describe the spiritual and ecological consequences of industrial contamination, drying springs, and climate disruption. The chapter foregrounds how land-based teachings, daily offerings, advocacy movements, and relational ethics sustain water protection across generations. It also shows how Indigenous water ceremonies integrate spiritual knowledge with environmental observation, creating holistic frameworks for resilience. Ultimately, the chapter reframes water governance as ceremonial, relational, and embodied, offering a decolonial model for ecological stewardship.

INTRODUCTION

As outlined in Chapter 1, ceremony functions as Indigenous governance, Indigenous water is not simply a physical necessity—it is a living, sacred relative in Indigenous worldviews, deserving reverence, protection, and ceremony. This chapter is a critical intervention into dominant environmental narratives by centering Indigenous water ceremonies as frameworks of ecological governance, cultural continuity, spiritual healing, and intergenerational resistance. Through critical thematic exploration, the chapter underlines how Indigenous Peoples mobilize spiritual practices, ancestral teachings, and daily ceremonial acts to resist settler colonial violence, restore water relationships, and reimagine sustainable futures rooted in relational accountability. This chapter addresses long-standing gaps in water governance liter-

DOI: 10.4018/979-8-3373-7559-5.ch003

ature by enriching Indigenous ceremonial knowledge systems—often marginalized or dismissed in mainstream environmental policy discourse—as sophisticated, alive systems of law, ethics, and environmental stewardship.

The chapter's primary objective is to explore how Indigenous water ceremonies resist settler colonial and industrial exploitation by asserting water as a sacred, sentient being essential to spiritual and ecological survival. This objective directly challenges Western frameworks that frame water as a commodity or "resource" and reframes water governance as a relational and ceremonial responsibility. Situating ceremonies as both acts of resistance and embodiments of Indigenous sovereignty, the chapter contributes to the growing field of Indigenous environmental justice. It brings particular attention to the gendered dimensions of water protection, especially the roles of women, grandmothers, and ceremonial leaders who have long defended muskegs, rivers, and wetlands through both ritual and political action. Another core aim of the chapter is to explore the integration of spiritual, ecological, and political practices in Indigenous water ceremonies as holistic systems of relational governance. This is achieved by examining how ceremonies—whether daily prayers, collective gatherings, or larger advocacy efforts—operate as systems of governance embedded in Indigenous worldviews. Far from being symbolic gestures, water ceremonies carry spiritual weight and ecological consequence. They function as spaces for education, mourning, healing, and organizing. For example, the chapter examines movements like *Keepers of the Water* in Canada that unite scientific monitoring, Elder guidance, and youth participation, modeling hybrid frameworks of Indigenous-led stewardship. In doing so, the chapter reveals how ceremonies uphold not only ecological balance but also ethical accountability, intergenerational learning, and decolonial governance (Norgaard, 2019).

A third major objective of the chapter is to verify the role of intergenerational teachings, daily ceremonial acts, and grandmother leadership in sustaining Indigenous water relationships as lifelong practices of cultural survival and spiritual integrity. This focus on everyday ceremony—silent prayers at a riverbank, morning offerings, respectful handling of water—demonstrates that ceremonial governance is not limited to formal gatherings but embedded in the rhythms of daily life. The chapter highlights how cultural memory and land-based childhood experiences connections to water, which evolve into lifelong responsibilities of care and protection. Grandmothers, in particular, emerge as spiritual anchors and knowledge keepers, guiding communities through personal transformation and ceremonial leadership. Their stories and presence affirm the continuity of Indigenous knowledge systems amid external disruption.

In addressing these objectives, Chapter Three makes several critical contributions to existing literature. First, it fills a persistent gap in the water governance and environmental policy fields where Indigenous ceremonies and ontologies are often excluded or tokenized. Many water frameworks tend to focus on technical, regulatory, or scientific dimensions, sidelining Indigenous knowledge as spiritual or anecdotal. This chapter refuses that binary, demonstrating that Indigenous ceremonies are comprehensive systems of science, law, and governance—complete and grounded in thousands of years of observation and practice. Following ceremonial knowledge, the chapter challenges the Eurocentric separation of spirit and science and affirms Indigenous environmentalism as both ancient and innovative. Second, the chapter addresses the underrepresentation of Indigenous women's voices in both environmental literature and governance structures. While much scholarly attention has been given to Indigenous resistance movements, there has been less focus on the ceremonial and gendered dimensions of environmental protection (LaDuke, 2020).

Centering the experiences, voices, and leadership of Indigenous women and grandmothers—who guide water ceremonies, carry ancestral warnings, and embody sacred responsibilities—this chapter foregrounds matriarchal governance as a cornerstone of environmental healing. It argues that Indigenous environmental leadership is inseparable from gendered roles and responsibilities passed through kinship and ceremonial lineage. Third, the chapter offers a significant reframing of urbanization and disconnection from land, which is often discussed in policy literature through the lens of infrastructure, migration, or modernization. Here, the urban environment is portrayed as spiritually barren—a place of disconnection from sacred waters, where concrete and metal replace the presence of life-giving rivers and springs. This narrative provides an important counter-perspective to dominant urban planning discourses and reasserts the necessity of spiritual relationships with water, even in modern contexts. The grief expressed by Indigenous Peoples living in urban areas reflects a deeper mourning of ecological and cultural rupture. Yet, the chapter shows that even in such spaces, ceremony continues—through acts as small as a whispered prayer or an intentional offering.

Structurally, the chapter is organized into thirteen overarching themes and nineteen subthemes that provide a nuanced, layered account of Indigenous relationships to water. These themes encompass a wide range of experiences, from environmental degradation and industrial extraction to advocacy, childhood memory, and ceremonial healing. This multi-thematic approach allows the chapter to hold complexity—honoring grief, joy, resistance, and renewal all at once. It creates a space where personal story, collective movement, spiritual guidance, and ecological analysis converge. Therefore, this chapter does not treat water ceremonies as past traditions in need of preservation; rather, it presents them as dynamic, evolving practices of resistance and regeneration. These ceremonies are future-making—they model sustainable relationships, collective responsibility, and spiritual resilience in the face of ongoing colonial violence and ecological collapse. As the chapter's final reflection emphasizes, healing is already happening: every grandmother who offers water to the river, every child who sings a water song, and every community that gathers in circle contributes to a wider movement of environmental and cultural restoration.

This chapter affirms that water ceremonies are not supplemental to Indigenous governance—they are central to it. Developing ecological protection in ceremonial practice, the chapter dismantles colonial separations between spirituality and policy, and offers an integrated, relational model of environmental healing. It contributes urgently needed perspectives to water governance literature by amplifying Indigenous voices, centering spiritual and gendered knowledges, and articulating ceremony as both everyday practice and political action. In a time of growing ecological crisis, this chapter provides not just analysis, but guidance—offering a path forward rooted in listening, relationship, and respect.

INDIGENOUS LAND-BASED WATER AS SACRED AND ESSENTIAL

Indigenous survival is inherently tied to sacred and protective relationships with land and water. These relationships are continually threatened by environmental destruction such as industrial extraction, pollution, and climate change. This theme explores the fundamental role that land and water play in sustaining Indigenous cultural identity, governance, and spiritual well-being. It underscores Indigenous Peoples' enduring responsibility to protect water-rich lands—such as muskegs, wetlands, and rivers—from extractive and harmful activities. Within this theme, two subthemes illustrate the dual experiences of active resistance and deep grief in response to water pollution and land degradation.

Protection of Water from Industrial and Environmental Threats

Indigenous communities have long engaged in resistance against industrial development as an act of ecological defense and cultural survival. Sacred water-bearing lands, especially muskegs and wetlands, are not just ecological zones—they are kin, spiritually alive and deserving of protection. This subtheme emphasizes how Indigenous women in particular lead efforts to protect these territories from strip mining and other extractive practices that threaten water systems. One illustrative statement from a community member reflects this grassroots leadership: "We work with a group of women that are noticing, trying to stop, strip mining of five area[s]. If those muskegs get strip-mined, it's going to create runoff into the rivers, streams, and lakes." The quote speaks to the embodied commitment and localized knowledge that Indigenous women bring to water protection. Ancestral teachings further stress the consequences of disturbing the land, reinforcing an ethic of ecological reciprocity that predates settler-colonial environmental frameworks. As one Elder shared, "The old people said that we shouldn't be taking things from the ground. Because if we take things from the ground, it will destroy everything around us." Such teachings articulate a longstanding cosmological understanding of balance and accountability between humans and the natural world. Through ceremonies, intergenerational storytelling, and collective action, Indigenous communities assert a water-centered ethic of responsibility. Protection of water is not only a spiritual obligation but also a sovereign act of governance and resistance to ongoing colonial disruption.

Challenges in Water Quality and Conservation

Colonial interventions in Indigenous water systems have drastically undermined both the quality of water and the trust Indigenous communities place in imposed water treatment systems. This subtheme explores how externally mandated practices—such as chlorination—have introduced new health risks and contributed to a breakdown in traditional water stewardship.

One community member notes the harmful health outcomes associated with state-controlled water treatment: "Our health department puts chlorine in our water to meet testing thresholds, then issues boil-water advisories. Many in our community refuse chlorinated water—diabetes, cancer, lung issues are rampant." The imposition of Western water governance methods—often without consultation or cultural consideration—has not only failed to resolve contamination issues but has introduced new layers of harm.

These challenges also carry an emotional and spiritual toll, expressed poignantly through intergenerational memory. A community Elder reflects, "When I was a child, if you cooked fish in a pot, you just cleaned the pot. Now you have to scrub it because the water is so dirty." This memory underscores the visible signs of degradation and mourning for water's lost purity.

In response, Indigenous water protectors have turned to community-led initiatives that aim to restore water quality and rebuild trust. These efforts often center on ceremonial practices, oral teachings, and land-based monitoring—reaffirming the interconnectedness between cultural survival and ecological renewal.

INDIGENOUS LAND-BASED CHALLENGES AS ENVIRONMENTAL DECLINE AND THE SACREDNESS OF WATER

Environmental collapse is not only a physical crisis—it is a spiritual rupture that threatens the balance between humans and the natural world. In Indigenous worldviews, water holds sacred status as a life-giving relative. The desecration of water systems—through drying springs, contamination, and climate disruption—signals not only ecological loss but a deep spiritual and ceremonial crisis. This theme explores how Indigenous communities interpret environmental change through spiritual, ceremonial, and ecological lenses.

Water as Life and the Blood of Mother Earth

Water is often described in Indigenous cosmology as the "blood of Mother Earth"—a sacred life force essential to both physical survival and spiritual wellness. This subtheme uses water metaphors to convey the profound grief and sickness associated with the loss of natural water sources. A powerful statement illustrates this understanding: "We used to drink straight from springs—our mother's blood. Now those springs have dried up. Fingers wither when blood stops, and so does the land." This poetic framing presents ecological harm as a spiritual crisis, likening the drying of springs to the cessation of blood flow in a living body. Such metaphors are not symbolic; they represent the lived experience of environmental collapse and its implications for cultural and ceremonial continuity.

This subtheme underscores that to damage water is to harm the entire living system of which humans are a part. It demands a spiritual and ceremonial response that acknowledges water as more than a utility—it is a conscious, feeling being that requires respect, gratitude, and protection.

Urgent Signs of Ecological Collapse

Indigenous communities are among the first to observe and interpret environmental warning signs due to generations of land-based knowledge, observation, and ceremony. This subtheme presents drying rivers, absent fish populations, and unpredictable seasonal cycles as indicators of a broader ecological emergency.

As one Indigenous representative shared, "Streams are drying up, the fish we used to see every spring haven't come for years." Such observations are not isolated anecdotes; they are collective expressions of ecological mourning grounded in centuries of environmental literacy and spiritual attentiveness. This type of monitoring—often dismissed by Western science as anecdotal—is deeply embedded in Indigenous ceremonial life. Observations of animal behavior, water flow, plant health, and seasonal rhythms are interpreted not only through empirical data but through stories, dreams, and ceremonial guidance. This holistic approach challenges the dominance of technocratic models of environmental assessment and affirms Indigenous science as a legitimate, rigorous, and relational knowledge system.

The urgency expressed in this subtheme is not apocalyptic, but prophetic. It is rooted in generations of warnings, teachings, and ceremonies that have long anticipated the consequences of ecological imbalance. It calls for a return to relational accountability—where water is not managed, but honored; not extracted, but engaged with as kin.

INDIGENOUS VIEWS OF URBAN DISCONNECTION AND ENVIRONMENTAL DEATH

Urban environments are often described by Indigenous Peoples as spiritually and ecologically lifeless. Unlike water-rich and ceremonial land-based settings, cities are viewed as places that sever human ties with sacred natural elements—particularly water. This theme explores the profound alienation Indigenous communities feel in urban landscapes dominated by concrete, metal, and industrial noise. It highlights the emotional, cultural, and spiritual dislocation that occurs when Indigenous peoples are removed from sacred water systems and the ceremonies that accompany them.

Cities as Places of Lifelessness

In many Indigenous perspectives, urban areas represent ecological and spiritual barrenness. Unlike traditional lands where rivers, springs, and muskegs are integrated into daily life and ceremonial rhythm, cities hide or degrade water sources. This subtheme highlights the grief that arises from disconnection. As one Indigenous speaker stated, "Now cities are of dead rock, metal and cement; nothing lives there." This sentiment captures the suffocating spiritual atmosphere experienced in urbanized spaces, where water is not accessible in its sacred form, but is instead filtered, chlorinated, and commodified. The disconnection from water in urban spaces symbolizes a broader cultural loss. Indigenous identity is deeply tied to place-based knowledge and the ceremonial honoring of water. In urban centers, the opportunities for daily offerings, water walks, or communal water ceremonies are limited or absent. Thus, cities are not merely inconvenient—they are perceived as existentially threatening to Indigenous spiritual and ecological well-being. The spiritual poverty of these spaces underscores the urgent need to return to water-based ceremonies that reestablish the sacred bond between people and waters.

Collective Responsibility and Advocacy

Indigenous water protection is not an individual task, but a communal responsibility rooted in ancestral teachings, ceremonial ethics, and governance structures.

This theme explores how Indigenous communities approach water care as a shared covenant—one that is activated through ceremony, advocacy, education, and relational stewardship. The subthemes address both organized communal action and the sacred power of personal rituals, revealing a multigenerational, multilayered framework for water protection.

Community Awareness and Action

Water governance within Indigenous communities is inherently collective, encompassing Elders, youth, leadership councils, educators, and ceremonial leaders. This subtheme focuses on how communities mobilize around water advocacy as a cultural and political responsibility. Governance is not solely administrative—it is ceremonial and relational. As a community representative stated, "It's not just four or five staff—our board, our Elders Council, it's everyone's responsibility to take care of water."

Such statements affirm that water care is an intergenerational and shared duty, embedded in the spiritual and political fabric of Indigenous life. Community gatherings, education programs, advocacy campaigns, and healing circles all contribute to a web of accountability. This model challenges Western environmental management frameworks that rely on technocratic and hierarchical governance, replacing them with relational and spiritual protocols. Water protection becomes not only a policy matter but a collective act of honoring life, memory, and future generations.

Everyday Acts of Prayer and Respect

This subtheme highlights how small, personal acts—silent prayers, morning offerings, respectful handling of water—constitute powerful expressions of spiritual governance. An Elder noted, "Even alone at the water's edge you can offer a silent prayer—water hears and responds." This quote underscores a foundational belief that water is sentient and responsive. Intention, presence, and humility activate the ceremonial bond. Everyday acts are not separate from formal ceremonies; they are micro-ceremonies that reaffirm relational ethics and spiritual integrity. Whether one is pouring water with care, singing softly to a stream, or remembering ancestral teachings while walking near a river, these acts function as continuous threads in the broader ceremonial fabric. They remind us that water protection is a daily commitment, not just a reaction to crisis. Respect for water must be habitual, humble, and rooted in love.

LAND-BASED WATER SUSTAINABILITY

For many Indigenous communities, relationships with water often begin in early childhood through emotional, sensory, and spiritual experiences. This theme explores how these formative interactions foster lifelong responsibilities to land, community, and ceremonial practice. Water is not introduced as an abstract concept but encountered as a being—one that teaches, heals, and inspires wonder.

Wonder, Emotion, and Formative Experiences

Childhood memories involving water are rich with curiosity and reverence. They serve as the emotional and spiritual foundation for later ceremonial responsibilities. One Indigenous voice recalled, "As a child I wondered why water looks different day vs. night—everything is a wonder." Such reflections speak to the spiritual resonance water holds even in the earliest stages of life. The visual beauty, reflective surface, sounds, and flow of water evoke awe and sacred attention.

These early impressions are more than personal anecdotes; they are part of a larger cultural process of transmission. Land-based childhood memory becomes the root of cultural continuity, where feelings of wonder evolve into practices of protection. Through these emotional and spiritual experiences, children begin to understand that water is not a passive element—it is kin, and with kinship comes responsibility. This subtheme affirms that ceremonial stewardship is not learned in isolation but nurtured through lived, embodied experiences from an early age.

Land-Based Balance, Reverence, and Sustainability

Indigenous ecological knowledge emphasizes balance, reciprocity, and gratitude. This theme explores how communities' express reverence for water through ceremony, song, and seasonal teachings. It presents gratitude as a guiding principle—one that ensures sustainable and respectful interactions with all elements of the natural world.

Learning Gratitude through Ceremonies and Songs

Gratitude ceremonies—often expressed through songs and offerings—are central to Indigenous environmental ethics. This subtheme emphasizes that spiritual expression and ecological sustainability are inseparable. A poignant quote from the chapter states, "We sing songs to water and all four elements—fire, sun, water, earth—in order to live in balance and express our thanks every day." These songs are not performative; they are prayers, ecological codes, and acknowledgments of interdependence.

Ceremonial gratitude aligns human behavior with the natural order. It fosters humility, attentiveness, and responsibility. Singing to water, individuals enter a reciprocal relationship, where they do not only take but also give. These practices challenge extractive worldviews that commodify water and promote overuse. Instead, gratitude rituals become acts of environmental justice— expressions of a worldview that honors balance, sustains ecosystems, and affirms Indigenous sovereignty.

The Water Keepers Movements

The *Keepers of the Water* movement exemplifies the merging of science, and collective action. This theme examines how this grassroots initiative emerged in response to environmental threats and continues to evolve as a spiritually grounded, community-led water protection network.

Formation in Response to Water Crisis

The *Keepers of the Water* movement was born from a ceremonial declaration that affirmed water as sacred. This subtheme details how this declaration catalyzed broader gatherings, advocacy efforts, and intergenerational education. As one participant recalled, "In 2006 we declared water sacred and formed *Keepers of the Water*. Gathering together—Elders, youth, scientists—is itself deeply healing." The movement bridges multiple ways of knowing, integrating Indigenous ceremony with Western scientific tools and community organizing. It provides a platform where diverse actors—grandmothers, youth, environmental monitors, and political advocates—converge around a shared ethic of water protection. What makes the *Keepers* unique is not only their effectiveness but their rootedness in ceremony. Their work affirms that healing water also means healing communities, restoring ceremony, and reclaiming Indigenous governance.

INDIGENOUS LAND-BASED HEALING AND CONNECTION

Ceremony is not merely a symbolic or spiritual act—it is a vital method of collective healing, relationship building, and cultural continuity. Indigenous water ceremonies serve as sacred gatherings where grief, memory, and renewal come together to foster deep connections among Indigenous and non-Indigenous peoples. This theme emphasizes how land-based gatherings supported in ceremony provide more than spiritual restoration—they also create a space for cultural transmission, community organizing, and cross-cultural solidarity.

Gathering as a Form of Healing

Water ceremonies are spaces where emotional release and communal healing occur without the need for words. Participants often report that simply gathering in a circle and offering collective intention is itself a transformative experience. As one participant reflected, "Just being able to stand in a circle, even in silence, and know we are offering to the water—that was enough." The quote reveals how presence, intention, and community activate healing. These moments of silence are not empty—they are sacred intervals filled with listening, honoring, and reciprocity. Ceremonial gatherings allow for grief to be expressed communally, particularly grief tied to environmental loss and cultural disruption. The circle becomes a ceremonial structure that holds multiple truths—mourning and hope, memory and transformation. In this way, gathering is a form of spiritual infrastructure where healing occurs not only within individuals but within relationships—between people, water, and land.

Relationship Building with Diverse Groups

Water ceremonies also create relational bridges across cultural, generational, and epistemological divides. Indigenous-led gatherings frequently include youth, Elders, and non-Indigenous allies, fostering networks of relational responsibility that expand beyond immediate communities. As one participant observed, "It's not just us—the youth, the allies, they show up to learn how to listen to the water too." This act of showing up, listening, and learning becomes connected in responsibility and reciprocal engagement. These moments of shared participation reframe environmental activism as relational and ceremonial, rather than adversarial or technocratic. Water gatherings do not only educate—they transform. Participants come to understand water not merely as an environmental issue, but as a relative deserving of care, presence, and ceremony. Such gatherings thus serve as powerful sites of decolonial education and environmental reconciliation.

INDIGENOUS LAND-BASED SACRED BEINGS AS WATER SUSTAINABILITY

Indigenous ceremonial practice is guided not only by Elders and community members, but also by sacred beings who inhabit the water realm. These spiritual figures—such as the mythical baby in the water—carry teachings about listening, humility, and responsibility. This theme highlights how ceremonial relationships with water are animated and guided by beings who speak through dreams, stories, and ritual practice.

Stories of the Baby in the Water

Stories of the baby in the water are more than mythology—they are spiritual protocols that remind communities to listen deeply and respectfully. As one Elder explained, "The baby in the water isn't just a story—it's a reminder that we must always listen when water speaks." This teaching reflects a broader epistemology in which spiritual beings are active guides, not metaphors. They carry lessons about humility, attentiveness, and care that are foundational to ceremonial life.

Elders are the human counterparts to these spiritual guides, ensuring that the ceremonial ethics of listening and respect are passed down across generations. Oral teachings, dreams, and ceremonies all serve as pathways through which the presence of water beings is acknowledged and honored. This subtheme affirms that ceremony is not just a human-led activity—it is co-created with the spiritual beings who inhabit the land and water.

Water Ceremony

Ceremonial leadership is not an assigned role, but an emergent responsibility grounded in family teachings, lived experience, and spiritual readiness. This theme explores how individuals are prepared to carry water teachings through mentorship, memory, and transformation. Becoming a ceremonial leader is both a personal and communal milestone.

Ceremonial Lineage

Many ceremonial leaders trace their knowledge back to teachings received from family members—parents, aunties, grandparents—who modeled protocols, prayers, and practices around water. One storyteller shared, "My dad and my auntie Diana—they were the ones who prepared me to carry this water work." Such relationships demonstrate that ceremonial leadership is rooted in kinship and obligation, not institutional authority. Family teachings ensure continuity. They pass on not only technical knowledge but also spiritual orientation—the ethics of humility, attentiveness, and relational accountability. Ceremonial lineage is thus a form of ancestral governance, where responsibility is inherited, not assumed.

Personal Transformation and Becoming a Kukum

For many, the journey into ceremonial leadership involves a deep personal transformation. Becoming a *kukum* (grandmother) is both a spiritual awakening and a community-recognized role. As one kukum reflected, "Two years ago I became a kukum—and I led my first water the day my son was born." This intersection of personal milestones and ceremonial initiation reveals how individual readiness and community trust converge in ceremonial roles.

Ceremony, then, is not a static set of rituals—it is a transformative path that involves emotional, spiritual, and relational preparation. It embodies nurturing wisdom, sacred leadership, and ceremonial strength, providing a model of how personal evolution feeds into communal wellness.

Indigenous Land-Based Connection to Water and Honoring Grandmother

Daily ceremonial acts, especially those centered around grandmothers and spiritual matriarchs, help to deepen the relationship between individuals and water. This theme illustrates how everyday practices—morning offerings, respectful conduct, remembering grandmother spirits—form the basis of a ceremonial ethic.

Daily Offerings and Personal Practice

Ceremonial relationships with water are renewed daily through personal rituals of gratitude, prayer, and reflection. These acts need not be public or formal to be powerful. As one Elder stated, "I make an offering to the water every morning—before I do anything else." This daily offering is a reaffirmation of relationship—a way of saying thank you, of being present, of engaging water as kin.

Through consistent offerings, individuals embody a long-term relationship of care and humility. These small acts sustain the spiritual infrastructure necessary for cultural and ecological resilience. They ensure that ceremony is not confined to events or gatherings but becomes integrated into the rhythms of daily life.

Respect and Everyday Ceremony

Ceremony is not confined to specific places or times—it exists in how people behave toward water in daily routines. One speaker emphasized, "Ceremony isn't only at gatherings. It's in how you pour, how you carry, how you thank." This quote underscores that water ceremony is embedded in everyday actions.

Respectful behavior toward water becomes an expression of ceremonial practice. Whether boiling water for tea or collecting rainwater, the way these acts are done—with reverence and presence—constitutes ceremony. These habits of care reflect a worldview where the sacred is ever-present and must be honored in all acts, no matter how mundane.

INDIGENOUS TRADITIONAL INNER HEALING AND WATER

Water has the capacity to facilitate profound emotional and spiritual healing. However, this healing is not automatic—it requires personal readiness and willingness to engage in emotional release. This theme explores the inner dimensions of water ceremony as pathways to self-cleansing and transformation.

Self-Cleansing and Emotional Readiness

Healing through water begins within. It demands vulnerability, openness, and emotional courage. One ceremonial guide shared, "You have to be ready. The water doesn't just take your pain—you have to let it go." This teaching reveals that water ceremony is an interactive process. The participant must be present—not only physically but emotionally and spiritually. Water listens and responds, but it does not act uninvited. Ceremony invites participants into a relational exchange where healing flows both ways. It is not about washing away pain, but about releasing it, trusting that water can hold and transform it. This healing is internal, yet it radiates outward, contributing to collective restoration.

Indigenous Land-Based Sacred Grandmothers and Ceremony Helpers

Grandmother spirits and ceremonial helpers form the spiritual backbone of Indigenous water governance. These feminine energies are not symbolic—they are guiding forces that ensure the integrity and continuity of ceremonial practices. This theme affirms the centrality of matriarchal knowledge in sustaining sacred relationships with water.

Grandmother Spirits and Their Roles

Grandmother spirits are seen as protectors, wisdom keepers, and guides who support prayer, shape ceremony, and model care. One Elder explained, "They say there are four generations—and each has helpers, just like we do." This teaching highlights the parallel between spiritual and earthly ceremonial structures. Just as community gatherings require helpers, so too do the spiritual realms.

Honoring grandmother spirits maintains the sacred architecture of ceremony. Their presence is felt in dreams, songs, rituals, and visions. They ensure that water ceremonies are not only technically correct but spiritually aligned. Their wisdom permeates every aspect of ceremonial life, offering strength, guidance, and balance across generations.

LEARNING REFLECTION

This chapter offers a powerful reimagining of water not as a resource, but as a sacred, sentient relative a being with whom Indigenous Peoples have maintained ceremonial, ethical, and ecological relationships for generations. Through the exploration of 13 themes and 19 subthemes, the chapter affirms that Indigenous water ceremonies are not only cultural or spiritual expressions; they are systems of governance, healing, and relational accountability. The knowledge and practices shared throughout this chapter reflect not only ancestral wisdom but also contemporary Indigenous resilience in the face of ongoing colonial, industrial, and environmental threats.

One of the most striking insights from this chapter is the consistent framing of water as an active participant in life. Water is not passive. It listens, remembers, and responds. This challenges dominant Euro-Western paradigms that reduce water to a commodity to be managed or extracted. Instead, Indigenous teachings reposition water as a teacher, healer, and relation—a being with whom one must be in constant ethical engagement. This reframing is central to the chapter's core objective: to document how water ceremonies serve as forms of both resistance and governance. Through prayers, offerings, songs, and stories, Indigenous communities reaffirm their responsibilities to water, and through water, to each other and to future generations.

From childhood memories of wonder to the formal responsibilities of becoming a ceremonial leader or *kukum*, the chapter illustrates how water relationships are sustained over the life course. These relationships are not just taught—they are lived, through daily offerings, respectful handling, and acts of gratitude. A grandmother making an offering at dawn, a child wondering at water's changing light, or an Elder singing to a stream—these are all acts of ceremony. This everyday ceremonial ethic is foundational to cultural continuity, demonstrating that ceremony is not only a formal event but a way of being that permeates ordinary life.

Many reflections within the chapter begin with grief: the drying of rivers, the contamination of sacred springs, the absence of fish that once marked seasonal renewal. These are not simply environmental losses—they are spiritual ruptures and cultural traumas. Urbanization, industrial expansion, and imposed water treatment systems have all contributed to a disconnection between people and water. Yet, this grief is not paralyzing. It is transformed into action—into song, advocacy, education, and the reaffirmation of Indigenous laws and values through ceremonial gatherings and water movements such as *Keepers of the Water*. These movements demonstrate that ceremony and activism are not opposites; they are interwoven practices of care and defense.

The chapter pays particular attention to the leadership of women and grandmothers in this work. Far from being symbolic figures, grandmothers are carriers of protocol, keepers of stories, and leaders in ceremonial preparation and intergenerational education. Their presence ensures continuity. The teaching of the four generations spirits and their helpers echoes a ceremonial architecture that extends across both visible and spiritual realms. These teachings reflect a matriarchal epistemology in which knowledge is not transmitted through institutional means but through relationships—through mentorship, storytelling, and lived example. These feminine knowledge systems are not supplementary to governance; they are governance.

This chapter also shows how Indigenous water ceremonies integrate science and spirit, observation and prayer. Community members are both ceremonial leaders and water stewards, using pH meters, seasonal monitoring, and traditional ecological knowledge to track changes and warn of danger. This hybrid approach defies colonial binaries. Rather than dividing spirit from science, Indigenous knowledge systems unite them through a holistic understanding of water's relationality. This synthesis strengthens both healing and governance, ensuring that environmental response is grounded in both data and cultural protocol.

Importantly, water ceremonies are also spaces for building bridges—between generations, communities, and cultures. The inclusion of youth and non-Indigenous allies in ceremonies reflects a growing commitment to relational responsibility and shared learning. Ceremonies become not only healing rituals but also pedagogical sites where people learn to listen to the land and to one another. This decolonial pedagogy offers a different model of environmental education—one rooted in humility, reciprocity, and presence.

Therefore, what this chapter demonstrates is that water ceremonies are not relics of the past. They are dynamic, evolving, future-making practices that model what it means to live ethically in a time of climate crisis, cultural fragmentation, and ecological decline. These ceremonies are forms of governance, forms of justice, and forms of healing. They do not look for salvation from outside—they enact it from within, through daily prayer, collective gatherings, intergenerational teachings, and spiritual relationships with land and water.

It becomes clear that healing is not only possible—it is already in motion. Every grandmother who prays by the river, every child who sings to the rain, every community that filters rainwater or protests a pipeline is participating in a sacred renewal. Healing happens in silence and in sound, in memory and in transformation. To follow water is to follow spirit, story, ceremony, and law. Water ceremonies remind us that the sacred is not separate from the political, that care is not separate from governance, and that survival is not separate from relationality.

This chapter invites us to understand that true sustainability is not technical—it is spiritual and relational. It is found in the humble acts of gratitude, the wisdom of Elders, the strength of matriarchs, and the resilience of communities who continue to honor water as life. Water ceremonies teach us not only how to protect the Earth but how to return to it—not as masters or managers, but as relatives, listeners, and caretakers.

This chapter has shown that Indigenous water ceremonies are not symbolic cultural expressions but systems of governance through which responsibilities to water, land, and future generations are enacted. Ceremony emerges here as a form of law—one that integrates spiritual accountability, ecological observation, and collective decision-making into everyday practice. By situating water as a sentient relative and ceremony as an ongoing ethical obligation, the chapter reframes governance as relational rather than administrative, grounded in matriarchal leadership, intergenerational teaching, and daily acts of care. These ceremonial frameworks do not stand apart from contemporary environmental challenges; they directly inform how communities respond to contamination, climate disruption, and imposed governance systems (Borrows, 2020). The governance practices articulated in this chapter provide the conceptual and ethical foundation for the next chapter, which turns to the material consequences of colonial water governance and industrial intervention. While Chapter 3 centers ceremony as the source of Indigenous law and authority, the following chapter examines how these governance systems are constrained, challenged, and resisted within contexts of environmental contamination, health impacts, and extractive economies. Read together, the chapters trace a movement from Indigenous law and ceremony to the structural conditions that make Indigenous-led water governance both urgently necessary and continually contested (LaDuke, 2020).

REFERENCES

Borrows, J. (2020). *Law's Indigenous Ethics*. University of Toronto Press.

LaDuke, W. (2020). *To Be a Water Protector: The Rise of the Wiindigoo Slayers*. Fernwood Publishing.

Norgaard, K. M. (2019). *Salmon and Acorns Feed Our People*. Rutgers University Press.

Chapter 4
Indigenous Water Stories:
Self-Determination

ABSTRACT

Chapter Four presents Indigenous water stories as a pathway to self-determination, emphasizing that water governance emerges from ceremony, emotion, memory, and land-based relationships rather than colonial policy structures. Through twenty-one themes and forty-two subthemes, the chapter reveals how Indigenous communities respond to water crisis, industrial contamination, climate disruption, and food insecurity with spiritual responsibility, technological innovation, and ancestral teachings. Water is portrayed as a sacred relative whose decline signifies both ecological and ceremonial rupture. The narratives highlight governance rooted in matriarchal leadership, grief work, land-based observation, renewable water initiatives, childhood memory, emergency preparedness, and embodied healing practices. Collectively, the stories demonstrate that Indigenous water sovereignty is enacted through daily offerings, community resistance, intergenerational knowledge exchange, and spiritual leadership such as kukum roles.

INTRODUCTION

Indigenous communities assert self-determination through their relationships with water systems. This chapter shows how ceremony, land-based knowledge, grief, climate observation, technological innovation, and memory come together to reclaim sovereignty over water and governance. Each theme in this chapter offers grounded stories from communities resisting colonial extraction and environmental destruction while simultaneously building futures rooted in ancestral values and spiritual vision. These stories reflect not only survival but also revival - revealing how water

DOI: 10.4018/979-8-3373-7559-5.ch004

cultural practices, sustainable infrastructure, traditional ecological knowledge, and oral teachings restore identity and reconfigure governance.

As outlined in Chapter 1, ceremony functions as Indigenous governance, this chapter is structured by narrating stories of toxic tailings ponds and climate grief and connecting the stories to land-based water harvesting, stories, water teaching, and grandmother-led ceremony. Indigenous peoples are not only victims of water-water disruption; they are architects of sovereign futures. Through this lens, water is no longer simply a commodity - it is a spiritual responsibility. The stories identify the role of governance as grown from relationship differentiating from imposition. The structure follows the logic of resurgence: first witnessing harm and grief, then highlighting innovation and healing. Together, the themes provide a powerful articulation of Indigenous water-water governance as both resistance and renewal. This chapter objectives include examining how Indigenous communities reclaim self-determination through relational governance of water systems; exploring the emotional, spiritual, and ecological dimensions of Indigenous land-based healing in response to water crisis, extraction, and state neglect; illuminating how water practices are carried out through ceremony, intergenerational teachings, and everyday acts of care.

Community concerns regarding illness and premature death in Fort Chipewyan are supported by epidemiological evidence documenting elevated cancer incidence in the region. A landmark health assessment by O'Neil et al. (2018) identified significantly higher-than-expected rates of several cancers among Fort Chipewyan residents when compared with provincial averages in Alberta. The study reported increased incidences of rare bile duct cancer (cholangiocarcinoma), as well as elevated rates of leukemia, lymphomas, and soft tissue cancers. While the authors cautioned against simplistic causal attribution, they emphasized that the observed cancer patterns were statistically unusual for a small population and warranted serious public health concern and continued investigation. For community members, these findings validated long-standing observations linking declining water quality, fish abnormalities, and industrial expansion to deteriorating health. This epidemiological evidence situates community testimony within a broader public health context, reinforcing that Indigenous knowledge of environmental harm is not anecdotal but aligned with scientific assessment of risk and exposure (Kelly et al. 2010).

INDIGENOUS LAND-BASED SURVIVAL AS SUSTAINABLE WATER

For Indigenous communities, land and water are not only sources of survival but sacred entities that require protection and reciprocity. This theme explores how environmental threats, water degradation, and traditional ecological knowledge shape Indigenous survival strategies. It frames land-based sustainability as a spiritual commitment, not just a practical one. The subthemes address industrial encroachment, water quality, and sacred teachings.

Protection of Land from Industrial and Environmental Threats

Industrial development is an existential threat to Indigenous lands, disrupting balance and sovereignty. This subtheme addresses how mining, extraction, and industrial expansion violate sacred relationships with the land and compel resistance from communities. In this regard, an Indigenous representative mention "we shouldn't be taking things from the ground. Because if we take things from the ground, it will destroy everything around us. And that is happening." This quote reflects the cumulative harm of industry and the call to defend land as a living being. Protecting land is an act of spiritual defense, not political convenience.

Challenges in Water Quality and Conservation

Water degradation directly threatens Indigenous health, culture, and ceremony. This subtheme discusses boil-water advisories, declining water levels, and pollution, showing how these disrupt daily life and spiritual practice. A pertinent quote from the story of an Indigenous individual is "you could drink water from anywhere. Now you can't do that. And we got busy when we moved into the towns to make a living, raise children. Suddenly, the water is not good enough to drink anymore. It all happened so fast. fast. Now, I mentioned this morning the springs, the natural springs we had, we used to go and get water from there. They're all dried up." The quote reveals the spiritual and practical consequences of unsafe water. Water quality is not a service issue - it is a sovereignty issue.

Traditional Ecological Knowledge and Sustainable Practices

Traditional ecological knowledge sustains life through observation, ceremony, and intergenerational teaching. This subtheme affirms that Indigenous ways of knowing about water cycles, weather, and wildlife form sustainable governance models. A relevant quote is "My grandmother said if the frogs stop singing, you better not plant yet. She was always right." which exemplifies traditional ecological knowledge as both empirical and ceremonial. Science and ceremony coexist in Indigenous environmental observation. The theme shows that land and water are sacred kin whose wellbeing is inseparable from community health. Resistance to industrial harm and conservation challenges are not acts of opposition, they are acts of protection and reverence. Traditional knowledge is not past tense rather it is the living guide for future sustainability.

CLIMATE CRISIS AND WATER SCARCITY

Water crisis has intensified water scarcity, altered ecosystems and threatened Indigenous livelihoods. This theme focuses on how communities experience the direct impacts of drought and ecological changes, particularly around major river systems like the Mackenzie. It shows that water scarcity is not abstract - it is a daily disruption to food systems, ceremony, and survival.

Drought and Water Crisis

The drying of the local water sources such as lacks marks a spiritual and ecological crisis. This subtheme discusses falling water levels, disrupted fish cycles, and spiritual grief around the loss of a living waterway. An Indigenous representative argues "The river, one of the biggest rivers in the world, they have never seen it as low as it is today. We are in a drought." reflecting both ecological disruption and emotional mourning for a river that is family. Water loss is not just environmental rather than a break in the water relationship. Water crisis for Indigenous communities is not a projection - it is present-tense. Dry riverbeds are not only ecological alarms; they are ceremonial wounds. This theme shows that Indigenous water stewardship must be central to climate responses.

INDIGENOUS LAND-BASED WATER AND HEALTH: INDUSTRIAL IMPACTS

Industrial extraction contaminates water sources and fuels public health crises in Indigenous communities. This theme explores the health consequences of bitumen extraction, tailings ponds, and polluted ecosystems - particularly in Fort Chip. It exposes the link between environmental harm and illness, especially cancer, and amplifies community demands for accountability (Kelly et al. 2010).

Bitumen Extraction and Water Use

Bitumen extraction requires vast amounts of water and leaves behind toxic waste. This subtheme discusses the volume of water withdrawn from rivers and the irreversible impacts on ecosystems and downstream communities. In this regard, an Indigenous individual states "So that project, it's one of the biggest projects on the planet. And so that project is here. The river flows right beside that project. It's not a project; there's like 30-some companies up there, and there's about 30-some projects up there. And so what they do is they dig for this bitumen – it's called open-pit mining." This quote underscores ecological disrespect embedded in industrial practice. Water cannot be extracted without ceremony and accountability as it is not infinite, and it is not inert.

Tailings Ponds and Contamination

Toxic tailings ponds leak into the water table, threatening food chains and family health. This subtheme highlights how carcinogens and heavy metals from tailings contaminate fish, birds, and drinking water, compounding fear and grief. A pertinent quote, "So, if you imagine these 90 ponds with this lake, liquid tailings ponds in Fort McMurray, you put them all together they are twice the size of the city of Vancouver." expresses the scale of natural systems with silent indication of unusefulness of water due to contamination. Contaminated water serves sacred relationships between humans, animals, and land.

Health Crises in Fort Chip

Fort Chip has become a frontline of industrial contamination, marked by cancer clusters and ecological crisis (O'Neil et al., 2018). This subtheme shares community testimony linking rare cancers to environmental exposure, and the sense of helplessness in the face of denial by industry and government. A related quote is "the water is making us sick because they get their water from the Athabasca River. And they have been saying the water is making us sick. And the people in Fort Chip are getting rare cancers." The quote captures the intersection of mourning and injustice, Health cannot be restored while denial persists - truth and water must flow together. This theme draws a direct line from extraction to illness. In the voices of Fort Chip and other communities, we hear not only suffering but clarity: water is not safe, cancer is not random, and justice is long overdue. These stories are not only data, but they are also testimony.

These findings underscore the need for enforceable environmental accountability and health-based water governance reforms, priorities taken up in Chapter 8's policy recommendations for governments and regulatory agencies. Scientific studies have also documented the presence of toxic contaminants in the Athabasca River system downstream of oil sands development, directly affecting the Fort Chipewyan region. Kelly et al. (2010) demonstrated that oil sands operations contribute measurable increases of mercury, arsenic, and polycyclic aromatic hydrocarbons (PAHs) into the aquatic environment through atmospheric deposition and surface runoff. These contaminants accumulate in sediments and bioaccumulate in fish and wildlife relied upon for subsistence, posing heightened risks to Indigenous communities whose diets remain closely tied to local waters. Importantly, the study challenged earlier claims that contamination originated solely from natural geological sources, instead identifying industrial activity as a significant contributor. For Fort Chipewyan residents, these findings corroborate observations of fish lesions, altered taste, and declining water trust. Environmental contamination, therefore, is not an abstract risk but a material disruption of food systems, cultural practices, and community health, reinforcing the inseparability of environmental integrity and Indigenous well-being.

LAND-BASED WATER SUPERVISION AND LOCAL PROTECTION

Water protection is increasingly led by community members using land-based knowledge and local resistance strategies. This theme focuses on how Indigenous individuals and communities take active roles in opposing harmful infrastructure and supporting water safety. It affirms that supervision and governance do not come solely from external authorities but through relational accountability.

Opposing Local Sewer Projects

Community members often resist infrastructure that threatens local ecosystems and water sources. This subtheme describes how residents mobilize against sewer projects that bypass Indigenous consultation and endanger sacred waterways. A pertinent quote is "Your sewer should never be by your lake, ever. Go put it in a different small pond or something; in Kikano, we have Whitefish Lake, and then we have some other little lake. That's where we put our sewer. You never put it in the water that you drink from because that's going to kill your lake." This quote emphasizes both the dismissal of Indigenous input and the environmental risk imposed by outside development. Resistance to local projects is not obstruction, it is defense rooted in long-term knowledge.

Supporting Communities in Action

Water protection requires solidarity - across generations, roles, and communities. This subtheme discusses how Elders, youth, and local leaders come together in ceremony and protest to safeguard water, showing how action is relational and collective. An Indigenous storyteller argues "We would come here and we would help bring people here and raise awareness so that happening. So that is a little bit of what we do." This quote highlights that activism and ceremony are intertwined. Protecting water is a community ceremony, where resistance becomes a sacred act. This theme shows how Indigenous water supervision is rooted in deep connection. It also shows that governance arises from relationships. Whether opposing a sewer project or organizing around ceremony, these stories show water protection as a sacred, shared duty.

Water Sovereignty and Self-Sufficiency

Water sovereignty begins with community-based self-sufficiency rooted in tradition and adaptation. This theme explores practical strategies like rainwater harvesting and highlights tensions between Indigenous methods and colonial practices. It frames water self-determination not only as resistance, but as care, creativity, and ancestral logic.

Rainwater Harvesting as Practical Empowerment

Rainwater collection is a land-based solution that reflects Indigenous environmental intelligence. This subtheme outlines how communities claim control over their water needs through harvesting systems, reducing reliance on unsafe or colonial infrastructure. A relevant quote is "By harvesting our own rainwater, by having systems on your own property so that you can always have access to water, is so empowering and so self-determining because it's one of those most important features, those most important aspects of our life is water." This quote affirms the practicality, autonomy, and empowerment of grassroots water infrastructure. Water self-sufficiency begins with simple tools, ancestral logic, and shared labor.

Resistance to Traditional and Indigenous Water Techniques

Indigenous water practices are often dismissed or obstructed by colonial systems, despite their proven value. This subtheme addresses how colonial regulation hinders Indigenous-led water solutions through bureaucracy, racism, or lack of recognition. An Indigenous representative argues "we have been researching but in trying to get it passed politically it's sometimes very difficult because we are trying to use things that are known and proven in the indigenous and in the old times but if you have somebody in a political sector that doesn't want you to go back to the way we used to do things there's resistance". This quote reveals the systemic bias that undermines Indigenous innovation. Bureaucracy must not be allowed to displace tradition and functionality.

Water sovereignty is not only a political concept but also a lived practice. From harvesting barrels to ceremonial teachings, Indigenous communities create sustainable futures by asserting the right to care for water in their own ways. This theme affirms that true self-determination begins in the backyard, not the boardroom.

WATER CRISIS AND HEALING THROUGH WATER

Water is a critical for pain, emotional release, and spiritual healing in Indigenous land-based practices. This theme highlights how water supports emotional processing and trauma healing. It discusses water as not only physical nourishment, but an energetic and spiritual medium for release and renewal.

Sustainable Water as a Pathway to Healing Emotional and Physical Trauma

Water is used to cleanse the body, mind, and spirit of trauma through gentle rituals. This subtheme describes how individuals engage with water through bathing, prayer, and presence to release grief, anxiety, and historical pain. A pertinent quote from an Indigenous storytelling is "The place where I go and sit and leave my tears. Every sickness in the body is an emotion that is not being let out. So, when we cry, that's that sadness, sadness, that loneliness, that hurt and that frustration that's leaving the body." This quote powerfully shows water as a receiver and cleanser of emotional burden. Water allows trauma to leave the body without words - just presence and trust.

This perspective supports the idea that water is a healer with agency. It does not ask for explanation or diagnosis. Instead, it invites release. Grief finds motion and voice in the river, the bath, and the offering. Through water, Indigenous communities process what has been silenced and rediscover a path back to feeling.

INDIGENOUS TRAUMA, MEMORY, AND SOMATIC HEALING

Water-based practices activate bodily memory and help release trauma stored in physical form. This theme focuses on the somatic dimension of trauma - how pain resides in muscles, nerves, and water patterns - and how land-based water rituals offer release through immersion, touch, and silence.

Trauma Stored in the Body and Released through Water Rituals

Ceremonial bathing and water offerings help move trauma through the body and out of it. This subtheme shares how people use water to reset their nervous systems and reclaim presence through ritual and sensory reconnection. In this regard, an Indigenous representative contends "You have to let that water carry you. You must mean it. You have to let go, and you have to let the water work on you, and you have to let the water heal you." This quote reveals how water ceremony reconnects body and breath, allowing healing from within. Healing is not always verbal, rather it can be somatic, slow, and felt through stillness with water.

This theme reminds us that trauma lives in the body long after the moment of harm. But water, in its rhythm, sound, and temperature, becomes a ceremonial partner in release. In Indigenous healing, the body is a site of memory and restoration, made sacred again through water rather than just a mere vessel.

INDIGENOUS LAND-BASED CHALLENGES AS ENVIRONMENTAL DECLINE OF WATER

Environmental decline threatens both the ecological balance and the sacred role of water in Indigenous lifeways. This theme highlights the dual realities of ecological collapse and spiritual responsibility. While land and water face increasing harm, Indigenous people continue to hold them sacred. The subthemes contrast reverence for water with the urgent signs of environmental crisis.

Water as Life and the Blood of Mother Earth

Water is not just essential for life - it is life itself, flowing as the sacred blood of the Earth. This subtheme frames water as kin, not resource. It explores how ceremonies and teachings position water as a living being with spirit and intelligence. Here an Indigenous storyteller asserts "If the blood in your hand is gone, I have seen a vision where the fingers are all withered because the blood stopped. And that is what is happening. With the land, we used to have springs coming down from these rocky hills we have along the Churchill River. This quote asserts a sacred worldview that links environmental destruction to cultural and spiritual harm. Treating water with ceremony is not symbolic - it is survival through reverence.

Urgent Signs of Ecological Collapse

Communities are witnessing the collapse of ecological systems that once sustained ceremonial and everyday life. This subtheme speaks to receding water levels, fish scarcity, drying wetlands, and the emotional weight of environmental change. An Indigenous speaker regrets by saying "One place close to Sandy Bay, a woman said she'd seen this little creek where there was fish. And the next morning, they said the fish were dead because the water had just evaporated or had just run out that same night. So, the fish didn't have a chance to get out of that creek." This quote expresses grief for lost places that once nourished body and spirit. Environmental collapse is a cultural and spiritual emergency, not just a scientific one.

This perspective calls us to understand water not only as an endangered resource, but as a sacred relative in crisis. When the land hurts, the people grieve. But they also act - through ceremony, resistance, and storytelling. In holding water sacred, Indigenous communities resist erasure and environmental apathy.

THE KEEPERS' MOVEMENT: ORIGINS AND GROWTH

The Keepers Movement began as a community-led response to water injustice and has grown into a spiritual and political force of Indigenous water protection. This theme traces the origin of the Keepers of the Water gatherings, focusing on their formation, goals, and ceremonial grounding. It shows how grassroots organizing and land-based spirituality converge to protect water and resist colonial governance.

Formation in Response to Water Crisis

The Water Keepers movement was born out of necessity - to defend sacred water and uplift Indigenous leadership. This subtheme explains how the gatherings began, how community members mobilized across regions, and how ceremony remained central to collective action. A related quote is "The water levels were dropping drastically, and the quality of the water was changing, and so they came together and said, 'We want to all work together to protect water.' So, they made a declaration in 2006, and that declaration, in summary, states that water is sacred and that we must work to protect it." This quote emphasizes autonomy and community-based authority in water governance. When governments fail, ceremony becomes governance and protection becomes a sacred agreement.

This perspective reminds us that movements don't begin with institutions - they begin with relationships. The Keepers Movement is more than advocacy - it is a reclamation of ceremonial law, led by those most intimately connected to the land and water.

INDIGENOUS COMMUNITY-LED WATER GOVERNANCE AND ACCESS

Access to clean water is not a technical issue - it is a matter of Indigenous governance, memory, and lived experience. This theme explores how personal experiences with illness, system failure, and cultural erasure inspire communities to create their own water governance models. It reveals that governance is shaped by memory, story, and responsibility - not policy alone.

Personal Experience and Health Impacts of Water Systems

When water systems fail, the consequences are not abstract - they are lived through illness and fear. This subtheme discusses chronic illnesses caused by poor water quality and how these experiences motivate individuals to pursue governance grounded in health and responsibility. A pertinent quote from an Indigenous speaker is "but for a long time I was drinking the water from the tap. And I have stomach issues due to that. Like, I got maybe parasites. So, I submitted that in my water claim as well, too." This quote highlights how lived experience and systemic neglect drive Indigenous-led solutions. Water governance begins with truth-telling about harm and healing.

Traditional Water Use and Cultural Memory

Memories of traditional water use guide current governance choices and ceremonial responsibilities. This subtheme explores how past practices - like carrying water or offering tobacco - remain embedded in modern systems of care and accountability. On this point, an Indigenous speaker argues "I still remember going to outhouses and then hauling our water from the lake and then straining the water using sheets, using like filters like that, like just normal sheets or shirts, clean shirts." This quote affirms that water treatment is not new - it is inherited and carried forward. Cultural memory is itself a learning tool, passed through teaching and practice.

Water treatment in Indigenous communities is not about creating something new - it's about remembering what already worked. Through personal health struggles and cultural teachings, this theme shows that solutions arise from lived experience, memory, and accountability to both water and community.

INDIGENOUS LAND-BASED PROSPECT AS WATER SOVEREIGNTY

Water sovereignty is limited by institutional barriers and colonial infrastructure - but grassroots innovation is growing. This theme explores the contradictions between federal or corporate water systems and Indigenous goals for self-determined infrastructure. It also shows how communities find creative ways to provide Water solutions, especially in emergencies.

Barriers to Expansion and Institutional Support

Efforts to expand Indigenous-led water systems are blocked by policies that ignore sovereignty. This subtheme outlines how grant requirements, regulatory frameworks, and jurisdictional conflicts limit Indigenous access to water infrastructure. A related quote is "we do have power outages. And it's not, like, right away that people can come fix it because of the distance. So, we have to sort of learn how to build capacity in a way. And then if there is a power outage that there's no panic." This quote reflects the structural exclusion that prevents communities from pursuing sustainable independence. True support for Indigenous water requires dismantling colonial eligibility frameworks.

Community-led Water Solutions and Preparedness

When institutions fail, Indigenous communities create their own emergency water systems from the land and kinship networks. This subtheme explores how communities prepare for power outages or fuel shortages by drawing on land-based skills, intergenerational knowledge, and local resources. In this respect, a quote from an Indigenous storytelling, "we do have the solar panels that we purchased when we got the global and water crisis grant. And then we do have the greenhouse to set up. And we started gardening as well too. It's all about learning to live off the land and, yeah. Great. So, like, community-led water, water governance, and the role of the women, how they're protecting the water." affirms self-sufficiency and community preparedness in the absence of state support. Water resilience is built from kinship, land-based knowledge, and ceremonial readiness, not corporate systems.

INDIGENOUS WATER AS SUSTAINABILITY AND INNOVATION

Indigenous communities are investing in renewable water that aligns with cultural values and environmental responsibility. This theme focuses on the ethical and sustainable aspects of Indigenous-led innovation, especially in solar and wind water. It positions clean water not as a trend but as a return to sacred responsibility.

Investing in Renewable Water

Renewable water projects reflect Indigenous values of non-extraction, long-term stewardship, and climate justice. This subtheme details how communities choose renewable options not only for efficiency, but to reflect values like humility, reciprocity, and reverence for the Earth. A pertinent quote is "What we were trying to do with our students in our camp was we were trying to invest in solar panels but then we could only go so far there because we bought the solar panels. It gives you an idea on how much consumption that we that we have, and that's why we invested in those solar panels." This quote connects technological decisions to spiritual and cultural logic. Clean water becomes sacred when guided by Indigenous teachings - not simply market trends.

This perspective reframes water innovation as a cultural and spiritual return. Investing in renewables is not assimilation - it is resurgence. By aligning technology with values, Indigenous communities are showing that sustainability is both ancestral and forward-looking.

Indigenous Community-Based Water Innovation

Innovation in Indigenous water governance is rooted in observation, community trust, and adaptability - not in imposed standards. This theme explores how Indigenous communities develop water solutions by watching, listening, and learning from their environment. It highlights local adaptations that arise through collaboration, rather than through top-down imposition.

Observing and Learning from Local Systems

Water innovation comes from watching how systems succeed or fail and adjusting with community wisdom. This subtheme explores how communities design or revise water systems by learning from their neighbors, adapting techniques to local land and needs, and trusting local knowledge. A related quote is "They have water collection system, and I think it's the biggest privately owned water collection system in Canada and it's just right beside us." This quote highlights intercommunity learning and the importance of land-based alignment in water planning. Innovation doesn't need invention rather it needs attention, listening, and trust in local land relationships.

This section shows that water innovation isn't always about new technology, it's about old relationships. By observing, adapting, and trusting one another, Indigenous communities create effective, sacred water solutions rooted in place, not protocol.

INDIGENOUS LAND-BASED CHALLENGES AS WATER CRISIS

Water crisis has disrupted food systems, animal patterns, and environmental cycles that are central to Indigenous life and ceremony. This theme explores how climate shifts erode traditional foodways and spiritual practices. It details changes to fish populations, water levels, and plant growth, while also capturing the cultural grief that comes with this disruption.

Impact of Industrial Influence on Indigenous Life

Industrial development accelerates water crisis and directly alters Indigenous ways of life. This subtheme discusses how roadways, pipelines, and extraction projects contribute to pollution, displace animals, and force lifestyle changes. An Indigenous storyteller asserts "We used to work in farms around here, my parents and myself. But after we married, the same thing. We go around to work; my husband feeds our kids. But at that time there is no liquor. But since that time, they opened the liquor to get the Indian people in the bar. It's worse. Now, they have started drinking. Even young people started drinking. Because they opened the liquor to buy a beer to go sit down drink beer in the bar." This quote reflects the layered impacts of industrial encroachment - not just ecological but spiritual. Climate damage is never just environmental, it is ceremonial, emotional, and structural.

Traditional Diets and the Loss of Clean, Natural Food

Food insecurity grows as traditional diets become harder to maintain due to ecosystem disruption. This subtheme explores how access to clean, hunted, and gathered foods has declined, leading to increased reliance on processed foods and the resulting health consequences. An Indigenous individual laments by saying "I remember. No sweet stuff. Just roger syrup. Long time ago. That's the one only. Nothing at all. Long time ago, we just ate meat and potatoes, fish, ducks, and most meat, their meat, rabbits, and ducks, and both chickens. We are so healthy bodies. All my family, we are so healthy. My brothers and my sisters." This quote expresses the shift from sacred nourishment to dependency, and the grief that follows. Food is not just fuel, it is identity, memory, and spiritual balance under threat.

Environmental Changes Affecting Water and Ecosystems

Fluctuating water levels and seasonal shifts challenge navigation, harvesting, and spiritual consistency. This subtheme speaks to how rivers, lakes, and wetlands behave unpredictably, impacting fish, plants, and the ability to hold seasonal ceremonies or travel safely. A related quote is "And then we had a sudden warm snap this spring. It melted all that snow. And because we were already drought the previous season, everything that melted was soaked up into the ground. It just absorbed. And our water is probably six feet lower than normal on our lake, Montreal Lake. And we have never seen that before." The quote captures the loss of guidance and rhythm once provided by predictable water behavior. Climate shifts silence sacred signals from the land, making ceremony and subsistence more uncertain.

Overall, this section demonstrates how water crisis strikes at the heart of Indigenous life - not just through science, but through ceremony, diet, movement, and emotion. The loss is not only ecological - it is cultural. Yet communities continue to adapt, remember, and resist disappearance.

INDIGENOUS LAND-BASED CHALLENGES FROM WATER CRISIS

Water crisis disrupts Indigenous lifeways by transforming weather, displacing animals, and threatening plant medicines essential to health and ceremony. This theme presents firsthand experiences of unpredictable weather patterns, disappearing food sources, and medicinal plant scarcity. It emphasizes that climate impacts are personal, spiritual, and deeply embodied. It contains seven subthemes reflecting the multi-layered consequences of environmental change.

Emergence of Flooding, Hurricanes, and Unpredictable Weather

Unusual weather patterns signal disruption and create danger in areas once considered stable. This subtheme shares how storms, floods, and hurricanes appear where they were once rare, and how this destabilizes land use, food access, and ceremony. An Indigenous speaker shared experience with the quote, "All I can remember basically growing up is there was a lot of snow. There was always deep, deep snow. And we didn't, you know, we never heard of hurricanes or flooding. And our lake water was always stable. There was never any flooding. But it was always deep snow. And we had our regular rain. It was regular rain and the regular snow." This quote captures the loss of predictability and relationship with once-familiar natural systems. When nature becomes unfamiliar, both safety and ceremony are put at risk.

Animal Migration and Declining Food Security

Animals once central to diet and tradition are no longer present, forcing communities into dependency. This subtheme outlines the disappearance or migration of species like caribou, beaver, or certain birds, and the resulting loss of food, fur, and ritual connection. An Indigenous speaker asserts "The only ones that I know is that we used to have a big population of moose, deer, and those ones are moving down south. They are migrating towards south. So, a big population of animals are migrating from north to south. That means you are having some food insecurity." The quote connects environmental change to emotional and cultural loss. Food scarcity caused by species loss is a rupture in memory and belonging.

Drying Water Bodies and Displacement of Wildlife

Rivers, streams, and ponds are disappearing - along with the plants and animals that rely on them. This subtheme addresses the shrinking of waterways and the cascading effects on hunting, trapping, fishing, and gathering. A related quote is "Small ponds, they are all drying up. Come October, September, October, they are all dry. In the past, there was always water there." This quote expresses the stark visual and ecological transformation taking place. When water disappears, the ecosystems and stories around it begin to fade too.

Fluctuating Lake Water Levels and Community Impact

Changes in lake depth and shape affect access, safety, and seasonal rhythms. This subtheme shows how families are unable to travel, fish, or hold ceremonies due to water unpredictability. A related quote is "It's really good for the community. It is good for the lake. It's very important. It's safe to drink when the water table comes up, it's cleaner, clean water and it provides for the community. It is good for the lake." This quote reflects the physical and emotional impact of water level on the community. Unstable water levels in lakes create a spiritual and logistical disconnect from place.

Water crisis Threatening Medicinal Plant Growth

Climate shifts interrupt the growth cycles and locations of sacred medicines. This subtheme highlights how plants traditionally used for medicine are harder to find, less potent, or disappearing entirely. A pertinent quote is "And then there's another one that we use, a rat root. It grows into those small lakes. And since those lakes are drying up, there's no more of that rat root. It doesn't grow there. It doesn't grow there anymore. Because it's dry." This quote communicates distress over changing relationships with healing plants. The disappearance of medicine is not only just ecological but also it is spiritual disorientation.

Trading Medicinal Plants Due to Scarcity

Communities must now exchange medicines across regions to meet ceremonial and health needs. This subtheme explains how certain medicines no longer grow locally, requiring intercommunity trade and adaptation. This perspective is underlined by an Indigenous speaker observation, "It's called rat root. It's getting to be scarce. And so what we are doing more is trading medicines. We are trading, different plants with different communities." This quote affirms relational adaptation rooted in ceremony and care. Climate resilience is collective - shared through reciprocity and ceremonial exchange.

Uncertainty in Food Sovereignty and Sustainability

Environmental unpredictability makes it harder to plan, store, or pass down food systems. This subtheme captures the fear and challenge of not knowing whether next year's land will offer what it has before. A related quote is "There is so much uncertainty with our climate now in terms of food security, food sovereignty, in terms of how we are going to sustain ourselves, in terms of sustainability. How are we going to feed ourselves if land is just drying up? either there's drought or either there's going to be flooding." This quote confronts the reality of challenging food sustainability. Sustainability is not only a concept anymore but also it is a question of whether culture can continue in a fast-changing world.

This section provides a sweeping account of how water crisis touches every part of Indigenous life: the animals, the medicines, the teachings, the ceremonies. While grief is ever-present, so is adaptation - through trade, observation, and intergenerational commitment. Land-based identity continues to evolve in response to Earth's transformations.

INDIGENOUS LAND-BASED KNOWLEDGE AS ENVIRONMENTAL MONITORING

Indigenous land-based knowledge complements scientific methods, offering detailed and relational approaches to monitoring environmental change. This theme contrasts Indigenous methods of water and food testing with state or industrial approaches. It demonstrates how safety, healing, and action come through lived, sensory, and spiritual experience.

Scientific Water Testing

Water testing blends modern science with traditional insight to safeguard community health. This subtheme discusses how communities conduct water monitoring by blending it with modern approach, often confirming what Elders have already observed. A related quote from an Indigenous storytelling, "So, part of our job was we tested the water and all the parameters that came back were good. Meaning the lake was not too acidic. It was not too much alkaline. The salt in the water was good. And the turbidity, the total dissolved solids, all those little things." Highlights the adoption of modern environmental monitoring. Science overlaps with land-based knowledge.

Safe Food and Shared Knowledge

Communities test and share food safety information as a form of collective care and resistance. This subtheme focuses on how people test and communicate information about fish, berries, and other traditional foods - sharing findings through ceremony, stories, and mutual respect. A related quote, "So, they pull their water for the village and everything from this lake. And of course, all the animals drink from it. And then they eat the fish from it. So, just so you know that all the food that you're going to eat is safe to eat." underlines relational trust as a core principle of environmental monitoring. Knowledge is not shared for recognition - but for protection and survival.

This theme affirms that monitoring the environment is not limited to instruments or degrees. It includes dreams, memories, taste, animal behavior, and ceremonial feeling. Indigenous science is grounded in trust, protection, and responsibility to future generations.

INDIGENOUS LAND-BASED CONNECTION TO THE WATER AND HONORING GRANDMOTHER

Water carries grandmother's teachings - ritualized in daily offerings, remembered through care, and honored as ongoing ceremony. This theme explores how the daily practice of water offerings becomes a way of remembering and continuing relationships with grandmothers and water spirits. It affirms that ceremony can be quiet, habitual, and deeply emotional.

Daily Offerings and Personal Practice

Pouring water and offering tobacco are acts of remembrance, not performance. This subtheme describes daily water offerings done in solitude or silence, grounded in gratitude, grief, and guidance. A related quote is "Ever since I was pregnant with him, I would go down to the lake, and I would smudge, and I would smudge, and I would sing the best way I knew how, for the water, for our grandmother, because she's a grandmother of ours." This quote highlights the personal and relational motivation behind daily ceremony treating water as grandmother. Daily practice is ceremony in motion - quiet, consistent, and powerful.

Respect and Everyday Ceremony

Ceremony does not need an audience rather it needs intention and respect. This subtheme discusses how care for water shows up in everyday acts - whether in how it's poured, stored, carried, or offered. A related quote is "Even if you're taking that, if you pour yourself a cup of water, breathe good words into her before you drink it, because she helps you as well." This quote affirms the continuation of ceremonial thought through the smallest actions. Every act with water is an opportunity to remember and re-honor teachings.

This perspective reminds us that water ceremony is not limited to public rituals. It lives in the home, in the kitchen, in memory of how a grandmother treated us. Through daily repetition, Indigenous women and their descendants turn habit into holiness.

INDIGENOUS TRADITIONAL INNER HEALING AND WATER

Water supports inner healing by making space for grief, spiritual renewal, and emotional release in private rituals. This theme explores water's role in cleansing the shelf - not only physically but emotionally and energetically. It reflects how self-directed healing requires readiness, solitude, and respect for water as a partner in emotional labor.

Self-Cleansing and Emotional Readiness

Healing requires readiness - and water answers when the spirit is open to receive it. This subtheme describes the ceremonial act of bathing, cleansing, or pouring water in solitude to prepare for emotional release and spiritual movement. A pertinent quote is "I wasn't in a good space and my waters weren't clean. But this is my own experience. So, as I started doing more of my own self-work, I've learned to acknowledge those waters within me." This quote affirms that healing is a collaborative act between self and water, rooted in timing and trust. Healing begins not with action, but with permission - from spirit, self, and water.

This perspective reminds us that water is always ready, but we may not be. Healing through water is not forced. It waits. It listens. And when the person is ready, it responds with gentleness, ceremony, and deep knowing. True healing is co-created.

INDIGENOUS LAND-BASED SPIRITUAL IDENTITY AND MEDICINE WORK

Water is central to spiritual identity and the emergence of medicine work through personal journey, naming, and relationship with the land. This theme explores how individuals find their spiritual path through water, receiving names, dreams, and responsibilities that define their healing role. It connects personal transformation with land-based spirituality.

Water in the Personal Healing Journey

Water awakens memory, spirit, and clarity during pivotal moments in a healer's journey. This subtheme describes how water functions as a mirror and guide, revealing inner truths that help shape identity and purpose. Such an experience is evident in a related quote, "my part started in 2007. So, I have been going to these meetings ever since, not every year, but most of the time. And to me, water is a very special, sacred element that we can use in our healing." of an Indigenous speaker. This quote honors water's role in healing and spiritual direction. Water reveals not what we ask, but what we need to remember.

Receiving the Spirit Name

Receiving a spirit name through ceremony near water connects a person to their medicine path. This subtheme highlights the power of naming through water, often during fast or spiritual encounters that mark transformation and acceptance of new responsibility. A pertinent quote is "When I asked for my spirit's name, I was called Medicine Spirit Woman. And I was really surprised because I thought it would have something to do with water. But you know, of course, water is medicine anyway. And then I was also told it has something to do with the little people." This quote captures the spiritual alignment that comes with naming and the water's role in confirmation. A spirit name is not just identity - it is instruction, gifted through water and witnessed by spirit.

The final section of the chapter affirms that spiritual identity is shaped by relationship - with water, with names, and with the land. Medicine work begins not with tools or titles, but with deep listening, readiness, and ceremony. When your spirit is named, you begin to walk with purpose.

LEARNING REFLECTION

This chapter demonstrates that Indigenous self-determination in relation to water is not solely a political aspiration, but a lived, ceremonial, and ethical practice grounded in daily relationships with land, memory, and kin. Across the narratives shared here, water emerges consistently as a living relative—one that carries grief, teaches responsibility, and sustains both bodily and spiritual wellbeing. These stories reveal that water governance is enacted not primarily through policy or infrastructure, but through ceremony, care, and relational accountability practiced over generations.

A central insight of this chapter is that environmental harm is experienced by Indigenous communities as a deeply embodied and emotional reality rather than an abstract ecological condition. In Fort Chipewyan, elevated cancer rates and declining water quality are not understood as isolated health events, but as outcomes of extractive systems that disrupt sacred water relationships. The convergence of community testimony with epidemiological and environmental evidence underscores that these harms are neither coincidental nor speculative; they are structural and ongoing. Yet, the chapter also foregrounds Indigenous responses that refuse victimhood. Communities turn to ceremony, water monitoring, rainwater harvesting, and grandmother-led teachings as pathways of resistance and renewal. The narratives further show that healing and governance are inseparable. Trauma, grief, and ecological loss are addressed through water-based practices that restore balance within the body and within community relationships. Leadership, particularly through kukum and matriarchal roles, emerges not through institutional appointment but through responsibility, readiness, and relational trust. This form of governance challenges colonial hierarchies by privileging humility, listening, and care over authority and control.

The chapter affirms that Indigenous water sovereignty is not achieved through confrontation alone, but through sustained presence, remembrance, and action. From everyday water offerings to community-led infrastructure initiatives, Indigenous peoples enact governance as a living practice. These experiences point forward to the policy pathways articulated in Chapter 8, demonstrating that just water futures must be grounded in Indigenous jurisdiction, ceremonial governance, and long-term relational accountability. The governance practices enacted across these stories—ceremonial leadership, local monitoring, renewable innovation, and community accountability—form the empirical and ethical basis for the policy pathways outlined in Chapter 8, which advocate for Indigenous jurisdiction, long-term infrastructure investment, and recognition of ceremonial governance as law.

REFERENCES

Kelly, E. N., Schindler, D. W., Hodson, P. V., Short, J. W., Radmanovich, R., & Nielsen, C. C. (2010). Oil sands development contributes elements toxic at low concentrations to the Athabasca River and its tributaries. *Proceedings of the National Academy of Sciences of the United States of America*, *107*(37), 16178–16183. DOI: 10.1073/pnas.1008754107 PMID: 20805486

O'Neil, A., Sojo, V., Fileborn, B., Scovelle, A. J., & Milner, A. (2018). The# MeToo movement: An opportunity in public health? *Lancet*, *391*(10140), 2587–2589. DOI: 10.1016/S0140-6736(18)30991-7 PMID: 30070210

Chapter 5
Indigenous Language as Water Governance

ABSTRACT

Chapter Five examines Indigenous language as a foundational system of ecological governance, cultural continuity, and water stewardship. Through community narratives, twelve themes, and twenty-seven subthemes, the chapter demonstrates that Indigenous languages function as living repositories of environmental ethics, ceremonial law, and intergenerational knowledge. It argues that language loss—driven by residential schools, colonial education, climate change, ecological disruption, and modernization—represents not only linguistic decline but the erosion of relational governance structures embedded in land-based worldviews. The chapter highlights how shifting animal migrations, heatwaves, and disrupted food systems diminish opportunities for contextual language use, thereby weakening cultural identity, discipline teachings, and spiritual memory. Yet, it also documents powerful Indigenous-led resurgence efforts: land-based immersion, Elder-guided teaching, cultural regalia, and family returns to ancestral territories.

INTRODUCTION

Indigenous water sovereignty is inseparable from Indigenous language, law, and land-based governance. This chapter argues that Indigenous languages are not cultural supplements to water policy but living governance systems that encode ethical responsibilities, ecological knowledge, and intergenerational accountability. Building on the evidence of contamination, health injustice, and community resistance presented in Chapter 4, this chapter centers how Indigenous governance is enacted through language, ceremony, discipline, and land-based education, and how these practices provide concrete pathways for water sovereignty. Indigenous languages carry precise knowledge about water flows, seasonal cycles, animal behaviour, and

DOI: 10.4018/979-8-3373-7559-5.ch005

environmental change. Words and relational expressions articulate how humans are expected to live with water rather than manage it as a resource. When these languages are disrupted through colonial schooling, environmental degradation, and displacement from land, water governance capacity is weakened. Water insecurity, therefore, is not only a technical or infrastructural problem; it is a governance crisis produced by the erosion of Indigenous linguistic and legal systems.

Colonial water governance frameworks continue to marginalize Indigenous authority by separating language from land and policy from lived practice. Water systems are regulated through technocratic institutions that privilege efficiency, control, and extractive development, while Indigenous governance is relegated to consultation or symbolic recognition. As demonstrated in Chapter 4, the consequences of this exclusion are visible in contamination, illness, and the breakdown of trust in water systems. These outcomes reveal that sustainability efforts which ignore Indigenous governance are structurally incomplete. Indigenous governance models operate through relational accountability rather than institutional hierarchy. Elders, grandmothers, and Knowledge Keepers guide decision-making through ceremony, teaching, and everyday practice. Discipline, regalia, and land-based instruction function as governance mechanisms, transmitting responsibility and authority across generations. These practices are not symbolic; they regulate behaviour, guide collective decision-making, and sustain water relationships. When Indigenous language is practiced on the land, water governance is enacted. As established in Chapter 1, our approach understands language as relational governance grounded in Indigenous law and ceremonial responsibility.

INDIGENOUS LAND-BASED RELATIONSHIP EXPRESSED THROUGH LANGUAGE

Indigenous languages are not simply linguistic systems of communication—they are land-based epistemologies and ontologies that reflect, sustain, and enact relationships between people, water, land, and all living beings. Language in this sense becomes a ceremonial, ecological, and legal expression of belonging. This section theme introduces the foundational understanding that language is an active agent in governing the relationship between Indigenous Peoples and their water and land systems. Through the act of speaking, language encodes and transmits responsibilities to protect, honor, and sustain the environment.

Language as a Bridge Between People and the Environment

Indigenous languages emerge from, and are shaped by, the natural world. Words for rivers, winds, lakes, plants, and animals are often descriptive, carrying embedded stories of origin, spiritual instructions, and ecological memory. In this way, language serves as a bridge between human and more-than-human worlds. One Indigenous knowledge holder shared, *"When we lose our Indigenous languages, we do lose the relationship that we have with each other as people, the priorities that we have, and the teachings of our people, because our language does promote a defined relationship with our environment."* This quote underscores that language is not passive—it is an active force of environmental and relational knowledge. Language in this context holds water memory: it tells us where the water flows, how it speaks, when to approach it, and how to give thanks. For instance, in Cree or Dene traditions, the naming of water bodies often involves ceremonial stories that caution against exploitation and reinforce relational accountability. Thus, losing language is not just a cultural loss; it is a severance from the teachings that guide sustainable living.

Language as a Tool for Environmental Stewardship

Indigenous languages function as systems of governance, wherein words are not merely descriptors but instructions (Hinton et al. (2018). Linguistic terms define seasonal changes, ceremonial practices, and ethical obligations for interacting with land and water. One Indigenous storyteller explained, *"I absolutely believe that our Indigenous languages are built with respect that ultimately does promote and strengthen a better climate around us."* This view highlights how environmental stewardship is embedded in the grammar and vocabulary of Indigenous languages. For example, specific terms for water creatures, sacred springs, or flood patterns often come with accompanying stories or taboos that regulate behavior to avoid imbalance or disrespect. The erosion of language threatens this embedded stewardship. Environmental governance that is rooted in Indigenous language can offer a counter-framework to settler colonial approaches (McIvor, 2020). Where settler law may define water in utilitarian terms, Indigenous language affirms its spirit, kinship, and life-giving role. Reviving language, therefore, is an act of reclaiming governance, responsibility, and climate justice (McCarty & Lee, 2014).

INDIGENOUS LAND-BASED IDENTITY PRESERVATION AS CULTURE AND LANGUAGE

Indigenous language does not only express ecological knowledge—it is also the heartbeat of identity, sovereignty, and cultural continuity. Language is how Indigenous Peoples remember who they are, where they come from, and to whom they are accountable, including to the land and waters. This theme focuses on how the preservation and revitalization of language is central to the resurgence of Indigenous cultural practices and self-determination. Language, in this sense, is a living archive of resilience, resistance, and survival.

Importance of Language in Cultural Continuity

Indigenous languages preserve intergenerational teachings, relationships, and responsibilities. Through language, oral traditions, spiritual teachings, and land-based protocols are passed down. These linguistic traditions carry the structure of sovereignty and the blueprint of collective identity. As one community member powerfully stated, *"Our sovereignty does not come from the government. The government does not give us sovereignty. Our language does because our language is what identifies us as a people."*

In water governance, this understanding is critical. Sovereignty over water is not just a political matter—it is also cultural and linguistic. Without the language to articulate Indigenous values, stories, and laws, the ability to self-govern and assert rights over water diminishes. Language revival, then, becomes both a cultural necessity and a political strategy. It ensures that water governance is not alien to community worldviews but emerges authentically from within them.

The Loss of Traditional Practices and the Need for Revival

Colonial disruptions—including residential schools, forced relocations, and environmental dispossession—have deeply affected the continuity of Indigenous languages and land-based practices. The loss of traditional activities such as trapping, fishing, and water ceremonies often leads to the erosion of the language that accompanied and gave meaning to those practices. One participant reflected, *"My dad would have the wildlife stretched, the coyotes, the wolves, the lynx. And I grew up like that... It saddens me that my grandchildren... the trapping ended with my dad."* This quote poignantly reveals how language and practice are interwoven—when one disappears, the other becomes vulnerable. Revitalizing language is not merely about recovering words—it is about restoring entire systems of cultural life and governance. Water governance, for example, cannot be decolonized without attending to the linguistic and ceremonial traditions that once governed how people interacted with rivers, lakes, and wetlands. Efforts to revive Indigenous language must therefore be deeply integrated with efforts to reclaim water sovereignty and environmental responsibility.

Indigenous language is both a cultural heritage and a governance tool for water and land. It carries the epistemological blueprints, ethical values, and spiritual instructions necessary for living in balance with the environment. As a medium of intergenerational knowledge and cultural identity, it sustains both the people and the ecosystems they are part of. Recognizing and supporting Indigenous language is essential not only for cultural preservation but also for transforming water governance into a more relational, responsible, and reciprocal practice grounded in Indigenous legal and ecological traditions.

INDIGENOUS LAND-BASED LANGUAGE THREATENED BY WATER CRISIS

Indigenous languages, deeply rooted in land-based relationships, are not just methods of communication but represent entire systems of knowledge, kinship, and environmental ethics. These languages carry land-based vocabularies, intergenerational stories, and cultural protocols that are tied to ecological rhythms, biotic presence, and traditional lifeways. However, water crisis is disrupting the very conditions that allow these languages to flourish. Melting ice, shifting seasons, declining biodiversity, and increasing weather volatility are not only threatening ecological systems—they are also accelerating the loss of Indigenous languages.

This theme explores the ecological dimensions of language endangerment, focusing on how water crisis affects Indigenous linguistic vitality through its impacts on the environment, traditional practices, and social coherence. Water does not operate in isolation—it threatens the social, ceremonial, and ecological contexts that sustain language use and transfer. In this sense, language extinction and climate collapse are intertwined crises. The subthemes below articulate how Indigenous language loss under climate change reflects not only the breakdown of ecosystems but also the rupture of cultural health and continuity.

Language as an Indicator of Cultural Health

The vitality of Indigenous language is often a mirror of cultural resilience. When a language is spoken fluently across generations, it is a sign of community health, robust intergenerational relationships, and land-based knowledge systems. Conversely, when a language is no longer spoken or passed down, it frequently indicates the weakening of ceremonial life, land-based practices, and cultural sovereignty. One Indigenous speaker articulated this connection powerfully: *"When you are looking at language as ontology, which means that language is a signifier of the way of the well-being of a culture. So if you have children who speak the language, supposedly that's a healthy culture of people."*

Language decline, in this sense, becomes a measurable symptom of broader cultural dislocation, particularly under the pressures of environmental upheaval. As climate change undermines the ecological conditions required for traditional practices, the opportunities to teach, learn, and speak the language in context also dwindle. Without the land, the language becomes disembodied.

Animal Migration as an Early Warning Sign

Many Indigenous languages contain rich lexicons associated with animals—each species tied to spiritual meaning, seasonal rhythms, and ecological teachings. As climate change alters migratory patterns or leads to the disappearance of species altogether, the words, stories, and ceremonial knowledge associated with these animals begin to vanish as well. One Indigenous participant observed, *"When you are looking at nature as, you know, struggling to provide. Like we, how do we know? We know by the migrations that are happening."* This quote highlights the role of animals as both ecological indicators and linguistic relatives. The loss of caribou, salmon, birds, or beavers not only harms biodiversity—it creates silences in language. Rituals, songs, and teachings that depend on these species fade into disuse, and with them, the vocabulary that carried place-based ecological ethics.

Limited Time for Cultural Communication due to Climate Effects

Climate change is reducing the physical and temporal spaces where cultural learning and language transmission traditionally take place. Extreme weather events, heatwaves, and unpredictable seasonal cycles reduce time spent on the land, which is often the primary site for informal teaching and storytelling. One community member remarked: *"Sometimes, when the heat wave hits... we have less time outside in nature to spend time in the water, to go traveling, just to be outside with each other. And therefore, you have less of an opportunity to just talk and be together."*

This observation reveals a key dynamic: language is often taught in action and in relationship. When people are confined indoors due to extreme weather or are forced to prioritize survival and adaptation over ceremony and leisure, the context for language transmission disappears. The relational space where language lives—at the water's edge, in the forest, around the fire—shrinks under climate duress.

Loss of Words due to Environmental Decline

Indigenous languages are rooted in specific geographies; the meanings of many words are intimately tied to species, landscapes, and ecological interactions. As climate change drives ecological loss, the relevance and remembrance of these terms begin to erode. An Indigenous storyteller shared, *"I don't even remember what it was called in our language. So that's a good example of the disappearance of language, because if you are not speaking it, you are not going to remember it."*

This reflection speaks to the grief of ecological and linguistic co-extinction. Words disappear not simply because they are forgotten, but because the world they describe no longer exists. The extinction of animals, plants, and habitats can lead to the extinction of language elements—entire epistemologies encoded in those words vanish. The loss is not only linguistic, but ontological.

Food Insecurity and Its Impact on Language and Culture

Traditional food systems are central to Indigenous language use. The processes of hunting, fishing, harvesting, and communal feasting are linguistic events—they are filled with teachings, protocols, and storytelling that reinforce cultural values and vocabulary. Climate change, however, is threatening the integrity of these food systems. Droughts, changing growing seasons, and species collapse are disrupting food sovereignty, and, by extension, the language embedded within food practices. An Indigenous voice stated: *"It's just a way of having that sacred respect for the food that we raise... that sense of gathering around the table as a family is a symbol of health... and it keeps the kids, and the family together."* As families lose the ability to engage in land-based food gathering and sharing, they also lose the shared cultural space where language is spoken, learned, and lived. If people stop doing, they stop saying. Indigenous language lives in the rhythms of everyday activities—not in abstraction but in action.

INDIGENOUS LAND-BASED LANGUAGE AS THE FOUNDATION OF WATER SUSTAINABILITY

Indigenous language revitalization is central to water sustainability because it restores the ethical frameworks through which water is governed. Language immersion camps, ceremonial gatherings, and land-based education reconnect learners to water systems in the ecological contexts from which knowledge emerges. These practices enable communities to observe environmental change, respond to risk, and uphold responsibilities to water as a living relation. Supporting Indigenous language revitalization is therefore a governance intervention rather than a cultural add-on.

Loss of Language as a Driver of Disconnection from Traditional Water

As Indigenous languages decline, so too does the worldview that positions humans as caretakers rather than dominators of water. When sacred words, ceremonial chants, and environmental protocols disappear, communities lose more than language—they lose the frameworks that guided ecological behavior. As one knowledge holder noted, *"The ceremonies that we have require the language. It requires the language as the chants are all in the language. The prayers are all in the language."* This points to a reciprocal cycle: environmental degradation disrupts ceremony, and the loss of ceremony weakens language. When language fades, so does the emotional, ethical, and spiritual connection to land. It becomes harder to feel responsible for that which one cannot name or speak to.

Contextual Language and the Impacts of Water Crisis

Indigenous languages are not universal—they are context-specific. Words are shaped by the environment they describe, by the snow that falls a certain way, by the wind from a particular direction, or by a season that brings certain foods. Water crisis disrupts these environmental contexts, rendering many linguistic elements obsolete or unused. As one community member shared: *"We are noticing that we've come away from land-based teachings so much that there are certain words that we just don't say anymore. Because they're only said in certain contexts."* This highlights the fragility of contextual language: it requires stable ecological and social patterns to thrive. As those patterns collapse, so does the relevance and use of the language. Words that depend on environmental phenomena become endangered when nature behaves unpredictably or ceases to behave in familiar ways.

Protecting Indigenous language is not only an act of cultural survival—it is a water strategy. These languages contain the principles of sustainability, relational accountability, and spiritual humility that are necessary to respond to the planetary crisis.

To protection Indigenous languages, we must also protect the lands, waters, species, and ceremonies they are rooted in. Language revival efforts must be land-based, community-driven, and intergenerational. Eventually, Indigenous languages do not only describe the world—they help sustain it.

Indigenous Land-Based Teaching as Sustainable Water

Indigenous land-based languages not only transmit words but also carry the embedded values, social rules, and moral teachings that define roles, responsibilities, and governance structures to protect water within a community. This theme foregrounds how traditional teachings about discipline, gender roles, and communal responsibilities were historically communicated through language rooted in land-based living. These teachings are being disrupted by modernization, climate change, and language loss, creating gaps in behavioral guidance and relational accountability, especially for younger generations. As discipline structures fade, so too does the everyday language that once taught respect, responsibility, and reciprocal living. This theme reveals that revitalizing Indigenous language must go together with restoring cultural systems of discipline and governance.

Disappearance of Discipline Teachings and Their Linguistic Foundations

Indigenous languages contain nuanced vocabulary for teaching discipline—terms and expressions designed not only to correct behavior but to cultivate respect for land, kin, and community. These teachings often emerged through indirect yet powerful forms, including storytelling, metaphor, and land-based analogies. However, many of these teachings are disappearing, as climate change and modernization reduce opportunities for intergenerational learning in traditional settings. One community member shared, "All the discipline language is associated with our ability to do work in and around the house. And so, if you have terrible weather that our kids aren't exposed to anymore, that entire knowledge base gets affected." Here, climate change becomes both an environmental and linguistic disruptor—limiting not only the physical engagement with land but also the verbal expressions that once accompanied such engagement. Without the conditions to perform traditional roles, the language of discipline no longer has a place to live.

The Impact of Modernization on Traditional Discipline

Modern schooling, digital technologies, and urbanized lifestyles are shifting the frameworks through which Indigenous youth understand authority and discipline. The relational teachings that once came through Elders, ceremonies, and time on the land are being replaced by institutional norms and screen-based influences. One Indigenous speaker notes, "Nobody chases their kids to go play outside anymore. And so, all of the play, idea of play, is affected. Play language. The idea of getting away from the phone, you know. And so, nobody wants their kids to be out in this heat." This observation reveals how climate conditions, urban fear, and technology have collapsed traditional spaces for play, which were also spaces for informal learning and respectful discipline. Disciplinary language was never abstract; it was modeled through everyday activity, embedded in land-based relationships. When these relationships are disrupted, the discipline language loses its grounding, and children become increasingly disconnected from the verbal and relational expectations that shaped prior generations.

This perspective eventually reveals that Indigenous language functions as a vessel of social governance and intergenerational authority. Without these teachings, there is not only linguistic loss but also a profound weakening of the social structures that sustained Indigenous life. Restoring discipline language requires restoring community-led, land-based education systems where these roles and responsibilities are once again spoken, demonstrated, and lived.

SPIRITUAL ENCOUNTERS AND MULTIGENERATIONAL MEMORY IN THE INDIGENOUS LANDSCAPE

Indigenous languages are deeply intertwined with the spiritual landscape and the multigenerational transmission of knowledge. This theme highlights how sacred stories and teachings—often told through land-based experiences—carry spiritual, ethical, and environmental wisdom across generations. These stories are not simply folklore; they are lived teachings, often rooted in ceremony and guided by visions or ancestral encounters. However, as language loss accelerates, so too does the community's ability to carry these stories forward with their full meaning intact. This theme argues that without language, the spiritual architecture of Indigenous worldviews is at risk.

Grandfather's Story and Its Message for the Youth

Elders carry knowledge that bridges the sacred and the everyday. Stories from grandfathers, grandmothers, and other knowledge-keepers often emerge from visions or ceremonial insight, and they are transmitted with specific language, cadence, and memory. One story was shared by an Elder who said, "The old man showed up, and becoming a ceremony, he said, 'They don't have to bring me home; there's nothing there, but I want my story to be told to the young people.'" This narrative conveys that spiritual responsibility doesn't end with death; it is transmitted across generations through storytelling. However, without the linguistic means to receive or retell these stories, their teachings are vulnerable to misinterpretation or neglect.

Language here is not just a tool of communication—it is the spiritual medium of transmission.

Language Barriers Between Ancestors and Present Generations

As fewer young people learn Indigenous languages, there is growing disconnection between the spiritual insights of past generations and the cultural understanding of the present. One representative poignantly reflected, "So we went, and we had a ceremony. I said, what did I do wrong? Nothing's coming. And the grandmother said, it's not you. It's us. We don't know how to fit the language for the people of this day to understand and how we, we talked in their time." This statement expresses the difficulty in translating ceremonial knowledge when the linguistic foundation has eroded. Elders may carry the stories, but without shared language, the youth cannot receive them. This spiritual gap reinforces the urgency of language preservation as a sacred responsibility.

This section demonstrates that Indigenous languages are not only spoken—they are experienced. They encode spiritual truths, ceremonial practices, and ancestral relationships. When these languages fall silent, the sacred memory of a people is endangered. Reclaiming language is thus not just about communication—it is about recovering Indigenous spiritual consciousness and multigenerational continuity.

INDIGENOUS LAND-BASED LANGUAGE LOSS AND ITS RECLAMATION

Although the suppression of Indigenous languages was systematic and violent, communities across Canada and beyond are actively reclaiming their linguistic heritage. This theme addresses both the trauma and healing associated with language revival. It explores how communities are resisting the colonial legacy of silencing by rebuilding fluency, re-establishing ceremony, and reclaiming everyday speech. Reclamation is not about nostalgia—it is about cultural survival, intergenerational strength, and sovereign expression.

Language Suppression in Residential Schools

The intergenerational trauma of residential schools has created silence and shame around speaking Indigenous languages. One community member shared, "What happened there is they didn't allow us to speak freely in our language. So, I lost my language when I was there." These schools were not simply institutions of education—they were instruments of linguistic genocide. The psychological impact of being punished for speaking one's mother tongue still reverberates across generations. The trauma is compounded by internalized shame, which often prevents survivors from teaching the language to their children, leading to further disconnection and cultural amnesia.

Importance of Speaking and Meeting Through Language

Despite this history, Indigenous peoples continue to fight for their languages with courage and creativity. Speaking, even imperfectly, is itself a political and spiritual act. As one participant noted, "They depend on the fish in the lake and the moose and the deer. And that's how they survived in those years." This quote, though modest in tone, expresses the deep interconnection between land, language, and survival. Reviving language restores relational worlds. Each word spoken is an act of return—to land, to ancestors, to collective memory.

This section affirms that the reclamation of Indigenous languages is a form of decolonial resistance and resurgence. It is a refusal to let colonial systems define the future. Communities are not just learning to speak again—they are healing, rebuilding, and re-establishing the ethical, spiritual, and ecological frameworks that their languages carry. Language revival is thus an act of cultural sovereignty, grounded in land-based practices, intergenerational love, and the refusal to forget.

INDIGENOUS LAND-BASED CULTURAL SYMBOLS AND REGALIA AS CARRIERS OF KNOWLEDGE

Indigenous cultural regalia and symbolic attire are not merely decorative artifacts or traditional garments; they constitute a sophisticated, embodied form of land-based language. This thematic focus draws attention to the ways in which Indigenous clothing functions as a living archive—a medium that transmits kinship roles, gender teachings, medicine knowledge, and governance protocols across generations. For many Indigenous communities, regalia is a visual articulation of story, ceremony, territory, and self-determination. This theme thus reframes regalia not as static heritage but as an active pedagogy—a method of teaching, remembering, and relating to both human and non-human relatives.

Traditional Clothing as the Language of Identity

Traditional regalia, gloves, beadwork, and other traditional garment's express identity in a way that often exceeds spoken language. In many Indigenous communities, attire signals one's familial lineage, regional affiliation, ceremonial responsibilities, and land-based relationships. As one participant shared, "So these are the representation that we have on our gloves, on our loved ones. And that's how you, we're telling people openly now, we're saying, hey, we know our medicines." This statement affirms how clothing and beadwork convey both spiritual and ecological teachings, embedding medicine knowledge in aesthetic form. These garments serve as relational markers—they speak on behalf of their wearers and situate them within ancestral, spiritual, and land-based frameworks. Regalia operates as a form of epistemological continuity. Unlike Western conceptions of clothing as individualistic fashion or utilitarian necessity, Indigenous garments are pedagogical and communal. The act of wearing them not only affirms one's identity but also maintains the lineage of knowledge associated with that identity (Simpson, 2017). In colonial contexts that sought to erase Indigenous presence through uniforms and imposed dress codes, the reemergence and public wearing of traditional attire constitute a powerful act of resurgence and relational affirmation (McCarty & Lee, 2014).

Indigenous regalia is also a material invocation of land-based teachings, particularly in how it draws from and reflects traditional medicines and environmental knowledge. Garments are often created using materials that come from specific territories—moss, animal hides, roots, shells, or medicinal dyes—all of which signify relationships with place. An Indigenous community member shared a transformative moment while fasting in Pine Ridge, Lakota territory: "We have such an abundance of medicine... I had no idea about this. But I went down to Lakota country... and I went to go fast over there." This quote illuminates the transnational relationships Indigenous peoples maintain with medicine lands and ceremonial places beyond settler-imposed borders. Garments sewn for ceremony or healing purposes become vessels of medicine, ceremony, and story. The making and wearing of regalia become a sacred encounter with place—a form of walking prayer or mobile ceremony (Tuck & McKenzie, 2015). When one wears clothing infused with materials from the land, one is not only wearing a piece of heritage but also carrying, remembering, and activating spiritual ecologies. The body becomes a site of reterritorialization, transforming daily practices of dressing and sewing into decolonial acts of reclamation.

This section affirms that Indigenous regalia is far more than cultural adornment—it is a land-based pedagogy, a living story, and a political-spiritual commitment to continuity. In contexts of displacement, cultural suppression, and linguistic erosion, these visual languages become essential mechanisms for survival and cultural regeneration.

INDIGENOUS FAMILY DISPLACEMENT AND INTERGENERATIONAL LANGUAGE LOSS

The ninth theme critically engages with how forced relocations and settler colonial interventions—particularly residential schools and state surveillance—have ruptured Indigenous families' capacity to transmit language, land knowledge, and cultural teachings intergenerationally. These displacements, though often enacted in efforts to protect children, simultaneously contributed to cultural erasure and linguistic fragmentation. Understanding this paradox is crucial to decolonizing narratives around "safety" and "education" that were weaponized to sever Indigenous connections to land and identity.

Relocation to Avoid Residential School

In the face of violent assimilation policies, some Indigenous families chose to relocate to remote or lesser-known areas in hopes of shielding their children from the reach of residential schools. While these acts were rooted in protective love and resistance, they nonetheless produced consequences for language transmission and ceremonial continuity. One strong testimony state:

My mother was born on the land in a place that in English is called Mira's Island... when she was 10, the RCMP were going to come to take her away... And my grandparents had enough of that... The older six already were taken off to residential school way before that.

This narrative reveals how families navigated impossible choices under colonial pressure. While some children were saved from residential institutions, the cost was often removal from cultural hubs and linguistic communities. Language loss thus emerged not only from direct educational violence but also from protective dislocation—an erasure enacted through the logics of survival.

Public Schooling and Cultural Erosion

Even when families avoided residential schools, their children were often funneled into state-controlled public education systems that functioned as assimilationist extensions of colonial policy. These institutions disregarded Indigenous languages and identities, compelling youth to learn and prioritize English. As a participant recounted, "They didn't have to go to residential school... but what they did have to do is go to public school... And so, the culture, the language... were still taken away." This quote underscores how public schooling was not a benign alternative, but a subtler mechanism of cultural suppression. This subtheme emphasizes how everyday policies—curriculum choices, language rules, and disciplinary structures—

systematically undermined Indigenous worldviews, promoting settler language as the sole path to legitimacy. Over time, the cumulative effect has been linguistic estrangement within families, where youth can no longer communicate fluently with Elders, losing access to oral stories, spiritual teachings, and land instructions.

Theme 9, therefore, situates intergenerational language loss as a layered phenomenon: one shaped by both overt violence and quieter displacements. Reclaiming this history is essential for reweaving the broken threads of land, language, and family.

INDIGENOUS LAND-BASED CULTURAL REVITALIZATION AS LAND AND LANGUAGE

While prior themes have explored the impacts of colonial disruption, Theme 10 turns to the ongoing revitalization efforts led by Indigenous families and communities who are returning to traditional lands to recover cultural memory and reawaken linguistic fluency. These acts of return are not symbolic but foundational: they represent a reconnection to ontologies, governance structures, and the spiritual relationalities that colonial systems tried to erase.

Returning to Traditional Territory

Returning to ancestral territory is both a political and spiritual gesture—a reclamation of space where memory and identity are encoded in landscape. As one participant explained, "So we've moved out to the traditional territory where my grandfathers lived in times past and have revitalized a lot of those teachings just as a family." This return is not just geographic; it is sensory, pedagogical, and ceremonial. On the land, families remember how to speak, how to gather, how to listen to plants, waters, and stories.

Land becomes the co-teacher—offering sounds, smells, and silences that awaken long-dormant knowledges. Unlike institutional programs that often isolate language from its ecological roots, land-based revitalization affirms that language and territory are co-constitutive.

Learning Language Through Resistance

Language revival, particularly in communities where fluency has been deeply fractured, is a courageous and often painful act of resistance. Many learners carry the shame, guilt, or grief of forgetting—imposed through years of colonial schooling or migration. One speaker stated: "I was heavily trained as a younger person to speak English properly... and they hyper-focused on making sure that I spoke

English properly. And I always regret that. Because what they took away from me was the natural ability to speak my languages."

This testimony speaks to the generational trauma and systemic pressures that stripped individuals of their linguistic inheritance. Yet, in choosing to learn, teach, and speak again—often in partnership with Elders and land—communities reject colonial narratives of disappearance. Speaking becomes an embodied refusal, a declaration of survival.

Theme 10 reframes revitalization not as a nostalgic recovery, but as an ongoing return—an active commitment to relationality, reciprocity, and responsibility. Language and land are not separate domains; they are intertwined vessels through which Indigenous futures are being rebuilt.

Theme 11, Indigenous Land-Based Language as Ancestral Code, shows that Indigenous languages function as ancestral codes—deeply encoded systems that carry not only speech, but ecological intelligence, kinship networks, and land-based cultural purpose. This theme explores how Indigenous languages are more than communicative tools; they are repositories of memory, place, and ethical responsibility. Each word spoken in Indigenous tongues is embedded with geographical, historical, and relational significance. It is not merely vocabulary that is at stake, but entire lifeways grounded in thousands of years of land-based living. Thus, learning and speaking the language is a radical act of reconnection—to territory, to ancestors, and to Indigenous laws of living. The following subthemes explore the interrelationship between land and language, as well as the emotional, spiritual, and cultural complexities of learning languages in displaced contexts.

The Role of Language in Land Connection

Indigenous language is not a detached grammatical system; it is an embodied way of being on and with the land. Each phrase, word, and syllable contain geographical intelligence, seasonal knowledge, and spiritual orientation. This subtheme focuses on the inextricable relationship between language and land. To speak in one's ancestral language is to locate oneself within a specific ecological and cultural map. Language expresses not only where one is from, but also how one belongs to that place and what responsibilities are entailed in that relationship.

One Indigenous storyteller powerfully reflects,

"At that time, we didn't have the language intact, my language. And so, we knew... that language is critical. It's like a code to our connection with the environment. It's our ancestors' knowledge... preserved over thousands of generations... all contained in our language."

This narrative illustrates that Indigenous languages are epistemological maps shaped by intergenerational relationships with the natural world. They are not learned solely for the sake of communication, but to reawaken the embodied memory and responsibilities toward land. In this sense, language is not a label affixed to nature, but a living method of relational engagement with the more-than-human world. Speaking one's language thus reaffirms the spiritual and ethical role one plays within their ancestral territory.

Urban Relocation for Language Access

Despite this deep connection between land and language, many Indigenous individuals and families find themselves relocating to urban areas to access language-learning opportunities and elder mentorship. This subtheme examines the difficult compromises made by those who must leave their territories in search of language fluency. In urban environments, institutional and community supports may be more accessible, but often at the cost of geographic disconnection from the very lands that shape the language being learned. An Indigenous representative describes this tension: "We moved back off the land and into the city, because that's where a lot of the elders in my area live, to be able to access supports and services for their health and well-being." This quote reflects how the colonial displacement of elders—due to health, infrastructure, and economic pressures—has in turn created a reversal where young people must now leave the land to find their knowledge keepers. While such moves can facilitate language access, they often do so in contexts far removed from the ecological roots of the language. The rhythms, metaphors, and teachings of the language are grounded in specific landscapes, seasonal changes, and traditional lifeways. When learning occurs without that environmental context, it risks becoming disconnected, abstracted, or commodified. As such, this theme highlights that for Indigenous language revitalization to be meaningful, it must be both linguistic and spatial. Finally, Indigenous language learning cannot be fully separated from land. Language must return to the territory that gives it life. Reconnection to both place and speech is necessary for cultural restoration to be holistic and sustainable.

INDIGENOUS ELDER WISDOM AND LANGUAGE IMMERSION

This theme centres Elders as sacred repositories of linguistic, cultural, and spiritual knowledge. Their teachings cannot be replicated through books or audio files; they are transmitted through lived, embodied relationships formed over time. Immersion in Elder-led environments creates the conditions for learners to absorb not just words but the ethical cadence, social obligations, and ceremonial depth embedded within language. As Elders grow older and pass on, the urgency to learn from them intensifies. Thus, language immersion with Elders is both a cultural practice and a race against time.

Gaining Fluency Through Elder Connection

The most profound language learning happens not in formal classroom settings, but through lived relationships with Elders who carry the language as an extension of themselves. This subtheme emphasizes how language acquisition deepens when embedded within everyday experiences—through storytelling, ceremony, gardening, fishing, or walking on the land. One community member reflects, "We connected with Elders and started to work towards fluency in Nuu-Tah-Nulth... I attained fluency and started to teach other people my language... in a land-based way." This statement affirms that fluency is not the endpoint but part of a generative process of reciprocity. Learning the language is not about accumulating knowledge but about entering into relational obligations—carrying forward the stories, protocols, and ethics that come with that knowledge. It is also about anchoring language in the land where it belongs. However, climate change and urban development are threatening these very landscapes, which may soon render certain expressions and narratives unintelligible. The urgency of immersion, then, is tied not only to the aging of Elders but also to the disappearance of the ecological conditions under which the language evolved.

Urgency of Preserving Elder Knowledge

The wisdom held by Elders is irreplaceable. Once they pass, entire worlds of language, culture, and spirituality may vanish with them. This subtheme underscores the importance of prioritizing Elder-led education and supporting their role as knowledge transmitters while they are still with us.

An Elder states with concern: "A lot of our experts were leaving us really quickly... Many of these Elders... would be gone a year or two later... That was the reality in my territory." This quote conveys the urgency and fragility of Indigenous language revitalization efforts. The passing of each Elder is not merely the loss of an individual but the silencing of a unique linguistic and cultural constellation. Therefore, immersion must happen now, while the wisdom still lives. It must be supported institutionally and culturally, ensuring that Elders are honoured not only as teachers but as living libraries and spiritual anchors of their nations.

LEARNING REFLECTION: RECLAIMING INDIGENOUS LANGUAGE AS LIVING RELATIONSHIP

This chapter demonstrates that Indigenous water sovereignty is sustained through language-based governance systems rather than institutional authority alone. Indigenous languages encode ethical responsibilities, ecological knowledge, and relational principles that guide how water is protected, shared, and respected. When these languages are practiced on the land, governance is enacted through ceremony, discipline, and intergenerational teaching. A central learning from this chapter is that water sustainability failures are not caused by a lack of technical knowledge, but by the systematic exclusion of Indigenous governance frameworks. Colonial water systems separate language from policy and ceremony from decision-making, undermining the very relationships that sustain water. In contrast, Indigenous governance operates through accountability to land, ancestors, and future generations. Language revitalization and land-based education therefore function as governance solutions, not cultural responses to crisis.

The chapter also highlights the role of Elders, particularly grandmothers, as water leaders whose authority is grounded in responsibility rather than position. Through ceremony, teaching, and everyday practice, they sustain water relationships and guide community decision-making. These forms of governance challenge colonial hierarchies by prioritizing humility, care, and relational responsibility. This chapter affirms that water governance is lived. Indigenous water sovereignty is enacted through daily practices that restore responsibility to water and reaffirm Indigenous law. These insights inform the policy pathways advanced in Chapter 8, which call for recognition of Indigenous jurisdiction, long-term support for land-based governance, and accountability to Indigenous legal orders.

REFERENCES

Hinton, L., Huss, L., & Roche, G. (Eds.). (2018). *The Routledge handbook of language revitalization*. Routledge. DOI: 10.4324/9781315561271

McCarty, T. L., & Lee, T. S. (2014). Critical culturally sustaining/revitalizing pedagogy and Indigenous education sovereignty. *Harvard Educational Review*, *84*(1), 101–124. DOI: 10.17763/haer.84.1.q83746nl5pj34216

McIvor, O. (2020). Indigenous language revitalization and applied linguistics: Parallel histories, shared futures? *Annual Review of Applied Linguistics*, *40*, 78–96. DOI: 10.1017/S0267190520000094

Simpson, L. B. (2017). *As we have always done: Indigenous freedom through radical resistance*. University of Minnesota Press. DOI: 10.5749/j.ctt1pwt77c

Tuck, E., & McKenzie, M. (2015). *Place in research: Theory, methodology, and methods*. Routledge.

Chapter 6
Indigenous Water as Healing and Resilience

ABSTRACT

This chapter positions Indigenous water healing as a holistic, relational, and intergenerational system of wellness that extends far beyond Western biomedical frameworks. Drawing from fifteen themes and thirty-six subthemes grounded in community stories, ceremonial teachings, and land-based practices, the chapter illustrates how water medicine encompasses emotional, spiritual, physical, and collective dimensions of health. It demonstrates how Elders' teachings on plant medicines, traditional foods, spiritual cleansing, and preventive care contribute to restoring balance disrupted by colonization, intergenerational trauma, capitalist pressures, and climate change. Water knowledge is presented as an ethical system rooted in ceremony, kinship, and ecological accountability. Healing emerges through reconnection—with land, community, identity, and ancestral teachings—whether through beading, listening practices in correctional institutions, or tending to ecologically damaged territories. The chapter ultimately reframes healing as a ceremonial, relational way of life.

INTRODUCTION

Despite growing scholarly attention to Indigenous wellness, the role of water healing as a full-spectrum system—addressing emotional, spiritual, physical, and collective wellness—remains significantly under-theorized in mainstream health literature. Much existing research tends to isolate either environmental relationships (Kimmerer, 2013), trauma recovery (Gone, 2013), or plant medicine (McGregor et al., 2020) without holistically incorporating them into Indigenous systems of care rooted in relationality, ceremony, and intergenerational storytelling. Additionally, health interventions for Indigenous communities are often still framed within Western biomedical models, which can reduce healing to pathology and treatment, excluding

Indigenous perspectives on balance, kinship, and sacred law (Hart, 2010; Waldram, 2008). This chapter addresses several key research gaps. First, while previous work has documented the significance of traditional medicine (Cook, 2014), few studies have explored the everyday incorporation of practices like water ceremonies as healing acts. Second, while some scholars have focused on intergenerational trauma (Truth and Reconciliation Commission of Canada, 2015), there remains a limited body of knowledge detailing the specific ceremonial, emotional, and relational methods Indigenous communities use to interrupt that trauma. This chapter builds upon foundational studies by Simpson (2011), Absolon (2011), and Green (2021), by centering Indigenous speakers who affirm that healing occurs through land, not despite it. Third, although Indigenous resurgence literature emphasizes land return and language revitalization (Corntassel, 2012; Tuck & Yang, 2012), little attention has been paid to the nuanced daily practices of water healing that occur in family homes, correctional systems, or climate-disrupted territories. This chapter fills that gap by offering stories from diverse community contexts, showing that healing does not require pristine environments—only sincerity and accountability.

Figure 1. How Cree First Nation water-energy sustainability is interconnected with community healing

Drawing on fifteen themes and thirty-six subthemes, this chapter expands current understandings of Indigenous health by illustrating how water healing is ongoing, ceremonial, and non-commodified. It challenges colonial and capitalist frameworks by foregrounding relational accountability, spiritual encounters, and collective care. In doing so, the chapter contributes an original, grounded account of how Indigenous communities are rebuilding healing systems through water knowledge, storytelling, and daily ritual.

INDIGENOUS WATER MEDICINE AS HOLISTIC HEALTH AND HEALING

Indigenous water medicine represents a holistic worldview of wellness grounded in the interconnectedness of body, mind, spirit, and relationships. Unlike Western biomedical models that tend to isolate symptoms and treatments, Indigenous healing emphasizes harmony and reciprocity across all domains of life. Health is not merely the absence of disease but the presence of balance—physical, emotional, spiritual, familial, and ecological. This theme explores how Indigenous communities practice healing through ancestral knowledge, plant-based relationships, oral teachings, ceremony, and intergenerational caregiving. The twelve subthemes presented below collectively illustrate how Indigenous medicine sustains individual and collective well-being through water living.

Water Medicine as a Communal Right, not a Commodity

Indigenous water healing traditions are rooted in collective responsibility and generosity. Medicine is viewed as a gift from the land and spirit world, not a product to be commodified. One Indigenous speaker asserted, "The medicines are for the kids, for the people, for your loved ones. They are not for sale. So, I don't agree with that, and I absolutely refuse to work with anybody who is selling medicine for profit." This position challenges the capitalist logic of privatizing knowledge and affirms that healing must remain embedded in relational accountability. Medicine belongs to the people and the land, not to individuals or markets.

Water as Blood and Immunity as Foundations of Healing

In Indigenous water knowledge systems, blood holds ancestral memory, spiritual power, and immunological strength. Health and disease are often interpreted through the lens of blood integrity. As one community member explained, "It's not so much about your immune system being weak. If the immune booster still doesn't do the job, the problem is always the blood. So, everything goes back to the blood as water." This understanding reflects an embodied water spirituality, where immunity is carried intergenerationally and must be nourished through both ceremony and physical care. Bloodlines are not only biological but spiritual pathways for wellness.

INDIGENOUS WATER MENTAL HEALTH AND CLIMATE CHANGE

Mental health challenges in Indigenous communities are tied to climate disruption, land disconnection, colonial systems. In order understand the water healing, we need to understand the mental structure of Indigenous community. This theme explores how mental distress is not solely internal - it is deeply relational. When the land is harmed, when families are displaced, and when traditional teachings are lost, emotional and psychological imbalance grows. Healing occurs through reconnection with ceremony, compassion, and traditional lifeways. The six subthemes in this section explore the social, ecological, and spiritual dimensions of mental wellness.

Mental Distress due to Disconnection from Land

Separation from ancestral land creates emotional imbalance and spiritual loss. An Indigenous Elder argues "Anytime you are disconnected from the land or any you can't be part of it, that affects your mental. You talk about people with addictions. Well, there's mental health issues there, too. Anytime they are removed from their natural environment, you know, that's going to affect them. The quote links a social crisis to land disconnection as a root of psychological suffering. Reconnection to land is necessary for mental healing and cultural continuity in Indigenous teaching.

Loss of Identity due to Colonization

Colonial systems fracture Indigenous identity and sever youth from their roots. This subtheme shows how identity struggles stem from imposed education, loss of language, and alienation from tradition. A relevant quote "We are seeing the extent of so many of our people, because of colonization, are so disconnected from their land and our way of life that they've forgotten who they are, loss of identity. And then that's affecting them." reflects how colonial imposition adversely affects Indigenous identity, contributing to self-doubt and dislocation. Water healing restores identity by replacing external definitions with ancestral truths.

Intergenerational Trauma and Genetic Impact

Colonial violence embeds itself in both memory and genetic inheritance. This subtheme explains how trauma is transmitted physically, emotionally, and spiritually across generations. A relevant quote, "genetically, we are being affected, too. Because trauma damages the brain and the heart and all of that. And so, we are seeing the extreme of all of that in our new generations. We are having that breakdown. Like, we've got more schizophrenics." conveys how trauma is embodied and passed through generations causing severe mental effect. Healing must work through both story and body to interrupt the cycle of inherited pain.

Healing Through Reconnection with Traditions

Traditional knowledge and cultural practices restore emotional clarity and spiritual belonging. This subtheme emphasizes that even small returns to traditional practices - like medicine picking - can revive mental health and reconnection. An indigenous representative argues "When I got sober, the one thing that helped me more than even going to rehab was going medicine picking. And I went on the Nipsey walk. And that was wonderful. Because it completely brought me back to myself. And that's something that no rehab, no other medication that they tried to put me on could ever do." This quote shows that even small acts of traditional practice revive mental clarity and emotional grounding. Cultural reconnection is a primary medicine for spiritual anxiety.

Healing Through Compassion and Understanding

Indigenous mental health care centers empathy and relational inquiry, not pathology. This subtheme contrasts Indigenous healing with Western diagnostic models, focusing on story, experience, and context. A related quote, "The kids went to me instead of going to the counselors. Because the counselors would hear these things that they are saying, get worried, and call the cops. Me, I'm not doing that stuff. So, whereas instead of going to the counselor, they were able to come to me, talk about their stuff, cry about their stuff, get through all those emotions that they cannot get through with modern medicine." reframes care through relational understanding, not punishment. Healing begins by honoring pain with understanding, not judgment.

Capitalism's Role in Youth Struggles

Capitalist systems alienate youth from land, identity, and community by promoting individualism and extraction. This subtheme identifies how exhaustion, burnout, and disconnection in Indigenous youth often result from chasing goals that conflict with relational worldviews. A pertinent quote from an Indigenous speaker is "The capitalism. Youths are the victims of capitalism. They want to make more profit." The quote critiques how capitalist systems contribute to disconnection and burnout among Indigenous youth. Water healing restores collective rhythms that resist capitalist speed and isolation.

Theme 2 teaches that mental health challenges in Indigenous communities are not just personal - they are environmental, historical, and systemic. Land loss, climate change, colonization, and capitalist values contribute to collective distress. Yet healing is found in reconnection - with land, ceremony, and one another. When young people return to traditional practices, when families listen without judgment, when communities gather around shared values, healing begins to unfold. Mental wellness, in this context, is not found in separation but in relational renewal.

INDIGENOUS WATER HEALING, INTERGENERATIONAL TRAUMA, AND RESILIENCE

Ceremony and cultural reconnection offer a pathway to healing intergenerational trauma caused by colonization and residential schools. This theme explores how Indigenous Peoples are restoring identity and spirit through renewed participation in ceremony. It focuses on how memory, language, and cultural strength are revived in healing settings. Through two subthemes, it addresses the damage caused by residential schools and the ceremonial rebuilding of self and collective strength.

Effects of Colonization and Residential Schools

Colonial systems attacked Indigenous spiritual strength by targeting ceremonial practices. This subtheme examines how residential schools were designed to suppress spiritual and cultural identity, with long-lasting psychological effects. An Indigenous representative argues "And then the schools destroyed that. My parents suffered just as much as we did in school. It wasn't so much we were a lot of kids suffered abuse, but for us it was the loneliness that just about killed us." identifying systematic oppression on the Indigenous heritage by the colonial residential schools. Disconnection from ceremony was part of colonial harm; restoring ceremony is now part of collective healing.

The Role of Ceremony in Healing and Strengthening Identity

Ceremonies reconnect individuals to community, identity, and spirit after generations of suppression. This subtheme highlights how sweat lodges, songs, language, and ritual practice support healing from trauma and reclaim cultural strength. In this regard, an Indigenous speaker quotes "So, there are ceremonies that go along with asking for rain to replenish and make healthy this land. So, all of the ceremonies that are life-giving ceremonies, they are healing ceremonies." This quote shows that ceremonial participation reconnects individuals with self and land. Ceremony helps individuals reclaim their place in Indigenous identity after disruption and displacement.

This section of the chapter shows that healing from intergenerational trauma cannot be achieved without restoring ceremony. Residential schools and colonial violence attempted to sever cultural ties by targeting rituals and language. But ceremony is how Indigenous people remember who they are and where they come from. By returning to water practices, individuals reclaim their spiritual and cultural power. Healing here is not only just recovery but also it is resistance, resurgence, and the rebuilding of intergenerational strength.

INDIGENOUS WATER SPIRITUAL CLEANSING AND PROTECTION

Spiritual cleansing and protective medicine practices help remove negative energy and restore balance. This theme highlights the protective uses of water medicine, specifically how plants are used to cleanse the body and ward off harmful energies. It includes one subtheme focused on wild rose root - a sacred plant used in ceremonial and spiritual defense.

Wild Rose Root as Medicine

Wild rose root is used in spiritual cleansing to protect against harmful energies. This subtheme reveals how the root of the wild rose plant is prepared and used in ceremonial cleansing to remove bad medicine from the body, whether caused by illness, jealousy, or spiritual interference. A related quote is "You boil it for about half an hour. Then you remove it and you soak your feet in there. So as soon as this person is done, soaking their feet for half an hour and praying, asking for that medicine to help them. And as soon as they're done, they have to take it outside and they say, "I give this to Mother Earth to take care of." This quote affirms wild rose as a spiritual tool that addresses energetic harm through physical ritual. Cleansing with wild rose root affirms the belief that not all illnesses are physical rather some must be addressed with spiritual medicine from the land.

This section demonstrates how spiritual wellness is maintained through respectful relationships with protective plants. Water medicine is not only about healing but also about defense - guarding the body, spirit, and community from harm. The wild rose root reminds us that the Earth provides protection from spiritual imbalance in addition to curing illness.

Indigenous Water Accountability as Knowledge

Indigenous healing practices are not merely systems of treatment—they are embedded in ceremonial ethics and spiritual accountability. This theme explores the sacred responsibility that healers carry, emphasizing that knowledge must be earned, stewarded, and shared with humility. Accountability within water healing traditions goes beyond technical know-how; it is rooted in ancestral trust, spiritual readiness, and relational obligations to the land and community. Knowledge, in this sense, is not a possession but a responsibility, and healing is not a commodity but a spiritual obligation.

Becoming a Teacher through Spirit-Led Encounters

In Indigenous worldviews, healing knowledge is received through spiritual instruction, not through self-promotion or formal certification. Those called to carry medicine often receive that calling through dreams, visions, or encounters with plant spirits. An Elder reflected, "Your job is to carry the medicine for the people. Your job is to travel all over and give them back that medicine. And to reteach them." This quote underscores the principle that medicine carriers are chosen by spirit and must be emotionally, mentally, and spiritually prepared for this role. The calling to heal is not a career path but a spiritual covenant—an agreement between healer, land, and community.

Healing Not for Profit

Healing must never be driven by profit. Many Elders speak forcefully against the commodification of Indigenous medicine, warning that when healing becomes a business, it loses its sacred power. As one speaker asserted, "If you want to sell medicine, don't come to my camp. You want to make a profit from medicine? Don't come around me. I'm not going to teach you anything. You want to disrespect and dishonor the medicine and take more than you need? One day that medicine won't like you." This quote emphasizes the relational nature of medicine—it requires respect, humility, and reciprocity. Medicine misused or taken out of context can lose its spirit, becoming not only ineffective but harmful.

Theme 5 reminds us that water healing is not just about knowledge transfer—it is about right relationship. Healing must be rooted in ethics, guided by spiritual readiness, and practiced in humility. Only then can medicine fulfill its sacred purpose.

INDIGENOUS WATER HAIR AND SKIN HEALING

Hair and skin are not merely surface-level concerns in Indigenous healing systems; they are conduits of spirit, memory, and self-esteem. This theme explores how water medicine nurtures hair and skin not only for physical restoration but for ceremonial strength, identity reclamation, and emotional well-being. These practices are intergenerational and often gendered, passed from Elders to youth as part of caring relationships and rites of passage.

Scalp Cleansing and Hair Growth

Hair holds sacred significance in many Indigenous traditions—it is a site of memory, prayer, and energy. Traditional scalp treatments are more than cosmetic; they are ceremonies of renewal. One community member described, "So you get that and you spray it on your head, and you rub your head upside down like this and you keep massaging your head. What it does is it clears out the follicles in the hair, and it unclogs them." This practice, often involving infusions of herbs and spiritual intention, reflects a holistic approach to healing that addresses physical health, identity, and spiritual balance simultaneously.

Moss and Skin Disorders

Mosses and other soft forest plants are used to treat skin irritations, infections, and other external ailments. These medicines are not merely practical—they are spiritually guided. One participant shared, "The spirits told me, go to the mountains and go get moss, take him to go get, and take him to the water in the mountain and wash him up." The act of gathering, preparing, and applying moss is a form of prayerful engagement with the land. Moss serves as a healing intermediary between body and spirit, providing both soothing touch and ceremonial presence.

Internal and Topical Healing Combination

Healing in Indigenous practice is systemic and integrated—what is applied externally must be matched with internal cleansing. One speaker explained, "So on my grandson, we also did the blood cleaner for him. So I did the blood, the kidney, the liver, and the blood cleaner for him. And he drank that every day. And we put the cream on every day." This disciplined, daily routine reflects an understanding that the body's systems are interconnected. Skin conditions are not isolated symptoms—they are signs of internal imbalance, to be treated with both plant-based infusions and ceremonial protocols.

Theme 6 illustrates that hair and skin care in Indigenous medicine reflects a commitment to dignity, ceremonial wholeness, and relational health. These treatments reconnect people to their bodies, their families, and the sacred nature of physical self-care.

WATER TRADITIONAL MEDICINES FOR PAIN AND INFLAMMATION

Pain, especially chronic pain related to aging or inflammation, is a daily reality for many. Indigenous healing addresses this not only with practical remedies, but with relational medicine—plants gathered with intention, prepared in ceremony, and applied with love. This theme highlights birch as a specific example of such a healing tool, demonstrating how water pain treatment weaves together physiology, spirituality, and environmental knowledge.

Birch Paper for Arthritis

Birch bark is a common, yet powerful medicine used for joint pain, inflammation, and arthritis. Its healing properties are activated through ceremonial preparation and precise application. As one Elder noted, "Somebody has arthritis on their spine. They would put that on their spine before they go to bed. You can put it on the spine, but it can't be put over the heart in the heart area. It should never be placed there." This teaching illustrates how Indigenous medicine involves exact knowledge—when, where, and how to apply each remedy. Birch is not simply applied; it is listened to, honored, and activated through relationship.

This theme reminds us that water healing is a living practice of care, guided by ancestral knowledge and present-day attentiveness. Pain is not only physical—it is spiritual and emotional, and birch offers comfort in each realm when used properly.

INDIGENOUS WATER INSTITUTIONAL HEALING AND COMPASSIONATE LISTENING

Healing must move beyond individual practice and enter the collective systems that shape daily life—schools, prisons, hospitals, and other institutions. This theme centers the role of listening as a decolonial and relational act within institutional spaces. Listening becomes a form of medicine, a cultural tool that reconnects those marginalized by carceral and medical systems with their own humanity and voice.

The Role of Listening in Corrections

Within prisons, Indigenous facilitators use water teachings to reintroduce relational practices like listening to incarcerated individuals. Listening becomes a sacred act, transforming spaces that otherwise enforce silence and punishment. As one speaker noted, "We don't want them around. And the problem is, they just needed to talk. And there was nobody to listen because of the shut up, shut up, shut up." This quote illustrates how disconnection and dehumanization result when people are denied the dignity of being heard.

Listening, within Indigenous frameworks, is more than passive hearing—it is a ceremonial practice that acknowledges the sacredness of another's story. Incarcerated individuals often carry layers of unspoken trauma. When someone listens to them without judgment, it becomes an act of witnessing, of relational validation, and of opening the doorway to healing. Compassionate listening becomes a water response to institutional harm.

This theme emphasizes that healing is incomplete unless it reaches into the systems that perpetuate disconnection. Bringing Indigenous values such as compassion and accountability into correctional settings interrupts the logic of erasure and affirms that every individual is worthy of restoration.

INDIGENOUS WAY OF HEALING THROUGH MEDICINE AND CEREMONY

Indigenous healing operates through a spiritual and ecological governance system. Medicine work is not casual or opportunistic—it is guided by sacred laws passed through generations. This theme reminds us that healing is rooted in ceremonial protocols that ensure the ethical, spiritual, and ecological integrity of all actions.

Gathering Protocols and Sacred Law

Healing ceremonies follow protocols that are dictated by land and spirit. Medicine must be gathered at specific times, in specific ways, with specific intentions. One Elder shared, "Wait till the seed pods are open... we wait for the seed pod to crack open and we shake them upside down." This protocol reflects a deep relational ethic: do not harm, do not rush, listen to the land's timing.

These protocols are about more than sustainability—they are about spiritual discipline. They ensure that healing remains rooted in reciprocity and humility, rather than extraction or commodification. When we follow these laws, we act in alignment with teachings that protect both people and ecosystems.

Indigenous Water Struggles and Lessons

Healing must also acknowledge the reality of ecological trauma. Colonization has not only impacted people—it has scarred the land. This theme explores how the land, like people, carries memory, injury, and the possibility for renewal.

Healing with the Scarred Land

The land is not always pristine; it carries its own trauma from clear-cutting, pollution, and forced displacement. As one participant stated, "What we realized was that the land was just as scarred as we are." This powerful realization affirms the parallel healing journeys of land and people.

To engage in healing, we must return to damaged land with care and ceremony. This return is not naive—it acknowledges pain while still choosing presence. In tending to be scarred land, we also tend to ourselves. Healing is not about erasure of history but about bearing witness to survival.

In conclusion, these areas reflect how Indigenous water practices root healing in relationship: to body, to story, to protocol, and to land. Whether through beading, listening, medicine, or returning to wounded places, healing is always ceremonial, always relational, and always ongoing.

INDIGENOUS WATER TRAUMA AND HEALING INTERRUPTION

Trauma in Indigenous communities is not merely an individual psychological condition—it is an embodied interruption in relational life, in ceremonial practice, and in the intergenerational flow of healing knowledge. Water trauma, whether resulting from colonization, displacement, residential schools, or systemic violence, affects not only the body but also the spirit and emotional presence of those impacted. Healing, therefore, must begin by addressing the ways trauma disrupts connection to land, ceremony, and kin.

Trauma Blocking Presence and Parent-Child Connection

Unresolved trauma often manifests as emotional withdrawal and spiritual absence, particularly within caregiving and parent-child relationships. Individuals may be physically present yet emotionally unreachable, unable to offer the warmth, attentiveness, and relational accountability that cultural teachings expect of caregivers. One Elder insightfully remarked, "That's what trauma does to you. It steals you away from your children and your grandchildren." This quote reveals the multidimensional cost of trauma—it not only injures individuals but also deprives children of essential teachings, ceremonies, and emotional grounding.

Water healing works to restore presence by inviting individuals back into relationship—with themselves, with others, and with the land. Ceremony, storytelling, plant medicine, and sensory experiences such as bathing and beading reawaken presence and help to mend the intergenerational fracture. As this theme reveals, healing is not simply doing—it is being. It is presence, reclaimed and restored.

WATER TRADITIONAL COMMUNITY HEALING AND SIGNS FROM NATURE

In Indigenous knowledge systems, the land is not a passive resource but an active, sentient teacher. The rhythms of weather, the behaviors of animals, and the presence of specific plants are all interpreted as signs—ceremonial messages guiding timing, action, and community readiness. This theme foregrounds how community healing is often sparked and confirmed through these ecological indicators, which are treated with reverence and relational intelligence.

Environmental Change and Spiritual Signs

Natural phenomena such as a bird landing, a thunderstorm approaching, or a sudden bloom of medicinal plants are not viewed as random—they are signs from the spirit world. These cues inform decisions about when to hold ceremony, what medicines to gather, or how to prepare for shifts in collective healing. One participant shared, "It was a crane, it just came and landed, so I was really thankful because for me I felt that it was a sign that our men are going to be starting to heal." This quote demonstrates the reciprocal listening that defines Indigenous healing. The land speaks—and communities listen.

Such practices challenge Western dichotomies between science and spirituality, between ecology and ceremony. Instead, Indigenous water healing merges these realms through relational attention and humility. Following the signs is not only practical—it is sacred.

WATER TRADITIONAL HEALING FOODS AND NUTRITIONAL MEDICINE

Indigenous foods are more than sources of nutrition—they are cultural memory, emotional medicine, and spiritual nourishment. Traditional diets, rooted in local ecologies and seasonal harvesting, carry knowledge about balance, wellness, and intergenerational caregiving. This theme highlights how water foods, especially plants like clover, play a significant role in holistic healing (Kimmerer, 2013; Norgaard, 2019).

Clover as Sweet Nutritional Support

Clover, with its soft sweetness and nourishing profile, is often used to support those who are emotionally depleted or recovering from illness. It is a comfort plant, ideal for children, Elders, or those needing gentleness in their healing journey. As one speaker described, "They'll dry all the clover. It tastes like honey. They're sweet. And especially when they're all ripe, they taste so good." This quote affirms the relational knowledge embedded in food—plants are not just ingredients, but kin with personality, intention, and emotional presence.

This food-based healing is relational, not prescriptive. It does not simply target symptoms but instead meets people where they are—offering kindness, memory, and sensory grounding. When shared in ceremony, even humble foods become acts of restoration, community-building, and cultural affirmation.

INDIGENOUS WATER PRACTICES FOR MEN'S HEALTH AND PROSTATE HEALING

Traditional Indigenous healing includes specific knowledge systems addressing men's health, including remedies for prostate care, spiritual balance, and emotional well-being. This theme demonstrates how men's healing is relational, rooted in land, and supported through gendered teachings passed down through ceremony and plant use. Healing for men involves returning to responsibilities—toward their own bodies, their communities, and the land.

Rabbit Root and Sage for Prostate Cancer

Rabbit root, often paired with sage, is traditionally used to treat inflammation and illness related to the prostate. The medicine not only addresses the physical condition but also reorients men toward ceremonial life. An Elder explained, "It looks like an umbrella, but it's also good for prostate cancer. Yes, and you would add it with sage, and that would heal prostate cancer." This quote emphasizes both the specificity and effectiveness of Indigenous plant medicine and its integration into a broader framework of accountability and care. This healing is not isolated from emotion, community, or gender roles—it is embedded in a spiritual economy of responsibility. Men are called not only to heal but to participate in collective well-being, returning to ceremony, to caregiving, and to the plants that sustain them.

Across these themes, Indigenous water healing reveals itself as an interwoven and relational system of care—responsive to trauma, guided by signs, nourished through food, and inclusive of all genders and generations. Healing is not compartmentalized; it is comprehensive. Whether addressing trauma's interruption of presence, listening for signs in nature, sharing sweet clover tea, or offering rabbit root for prostate health, these practices speak to a sacred order where wellness is sustained through humility, ceremony, and kinship with the land. Together, these teachings reaffirm that healing is not a service—it is a way of living in relationship.

LEARNING REFLECTION: RESTORING HEALING AS RELATIONAL CEREMONY

This chapter, "Healing and Resilience: Stories of Water Connections," offers a transformative redefinition of what it means to heal. Rather than approaching healing as a compartmentalized, clinical practice, this chapter affirms that Indigenous water healing is a full cosmology—a system of care that is simultaneously physical, emotional, spiritual, environmental, and political. Across fifteen themes and thirty-six subthemes, it becomes evident that healing is not episodic or reactionary, but cyclical, collective, and continuous. It is not only about the body, but about memory, presence, belonging, and ceremony. It is not only about individuals, but about land, family, and community.

One of the foundational teachings to emerge from this chapter is the idea that land is not a setting where healing occurs, but the very medicine itself. Land is the healer. Through plants, roots, moss, water, and weather, the Earth provides tools for recovery and renewal—but not in ways that align with Western medical paradigms. These medicines are not passive or abstract; they are active relatives, imbued with spirit, carrying stories, memory, and instructions. As Indigenous Elders remind us throughout the text, these medicines cannot be commodified or separated from their relational and ceremonial responsibilities. As shared in Theme 1, "Medicine as a Communal Right, Not a Commodity," medicine loses its efficacy when separated from humility and relational ethics.

The chapter also teaches that healing is not always synonymous with curing. Theme after theme reinforces that healing is a return—return to land, to ceremony, to community, and to body. Healing is about reconnecting with the self and with others after the violence of colonial interruption. This is most powerfully illustrated in Theme 3, "Intergenerational Trauma and Resilience," where ceremonial participation—such as sweat lodges, medicine picking, and language revival—supports identity restoration for those who have survived residential schools and colonial erasure.

Importantly, the chapter challenges the reader to reconsider trauma as more than a personal wound. As Themes 2 and 12 argue, trauma is environmental, historical, and relational. It is inherited across generations and exacerbated by disconnection from land, ceremony, and relational care. Yet the chapter also shows that resilience is equally intergenerational. Healing is not about forgetting pain but learning how to walk alongside it with ceremony and memory. As one Elder reflects, healing happens even on clear-cut land. It does not require perfection—it requires sincerity, intention, and accountability.

This reflection cannot overlook the sensory and embodied nature of Indigenous medicine. Themes on sun-infused tea, moss wrappings, and clover drinks teach that healing moves through the senses. Cedar, clover, birch, and wild rose are not just physical substances but are also pedagogical tools. They teach warmth, rhythm, patience, and gratitude. As noted in Theme 1.7, "Restoring Internal Warmth Through Sun-Infused Medicine," the healing warmth of maple bud tea teaches that care is not always dramatic or immediate—it can be seasonal, slow, and quiet. These healing practices demand listening—not only to human stories but also to the language of plants, wind, water, and silence.

Water teachings throughout the chapter also offer profound insights. Water healing is ceremonial ways that go beyond gender binary roles. Certain teachings—like prostate health, womb protection, water carrying, and food preparation—are carried by specific relations: grandmothers, aunties, uncles, and fathers. These roles are not assigned by institutions but earned through spirit encounters and ancestral responsibilities. Gender, in this worldview, is sacred, relational, and ceremonial—not rigid or hierarchical. Crucially, the chapter emphasizes that healing is not confined to rural or pristine environments. It occurs in correctional institutions, urban homes, and damaged ecosystems. In Theme 9, "Institutional Healing and Compassionate

Listening," the act of listening within prisons is framed as ceremony. Offering space for stories, without judgment or surveillance, becomes a radical act of healing. Similarly, Theme 11, "Healing with the Scarred Land," reminds us that damaged land still holds memory and power. Healing does not require a return to an unbroken past—it requires presence and participation, even amid loss.

The role of food, ceremony, and compassionate care as forms of healing also surfaces in multiple themes. Clover tea, discussed in Theme 14, is not only nourishing but also comforting—sweet, simple, and emotionally grounding. Similarly, beading, as explored in Theme 8, becomes a method of emotional processing and ancestral communication. The rhythm of each stitch is a grounding force, a prayer, and an invitation to feel stress without being consumed by it. One of the most transformative teachings from this chapter is the understanding of signs and messages from the Earth. As illustrated in Theme 13, animals, weather, and environmental changes offer ceremonial instructions and warnings. Healing, then, is not a silent process—it is a dialogue between land and people. To walk a healing path, one must listen deeply to the Earth and recognize when the land speaks.

Perhaps the most profound lesson is this: healing is not a program to be implemented or a protocol to be prescribed. It is a daily practice of remembering what is sacred, acting with humility, and maintaining balance with all relations—human and non-human. Healing is carried through moss and moon cycles, beading and birch bark, silence and song. It is preserved through ceremony, intuition, storytelling, and touch. This chapter does not offer instructions—it offers an invitation. An invitation to return to land, to remember, and to walk again with integrity, love, and purpose. In this way, Indigenous water healing is not a supplemental practice. It is a sovereign, complete, and intergenerational healing system—built from relationships, sustained by spirit, and guided by the land itself. Healing, as this chapter teaches, is not only about fixing what is broken—it is about remembering what has always been sacred. And in remembering, we begin to restore. These ceremonial teachings are explored in depth in Chapter 7.

REFERENCES

Absolon, K. (2011). *Kaandossiwin: How we come to know*. Fernwood Publishing.

Cook, S. (2014). *The dancing plants: Healing medicines and Indigenous knowledge*. University of Manitoba Press.

Corntassel, J. (2012). Re-envisioning resurgence: Indigenous pathways to decolonization and sustainable self-determination. *Decolonization, 1*(1), 86–101.

Gone, J. P. (2013). Redressing First Nations historical trauma: Theorizing mechanisms for Indigenous culture as mental health treatment. *Transcultural Psychiatry, 50*(5), 683–706. DOI: 10.1177/1363461513487669 PMID: 23715822

Green, J. (2021). *Indigenous health and the politics of healing*. University of Toronto Press.

Hart, M. A. (2010). Indigenous worldviews, knowledge, and research: The development of an Indigenous research paradigm. *Journal of Indigenous Voices in Social Work, 1*(1), 1–16.

Kimmerer, R. W. (2013). *Braiding sweetgrass: Indigenous wisdom, scientific knowledge, and the teachings of plants*. Milkweed Editions.

McGregor, D., Restoule, J. P., & Johnston, R. (2020). *Indigenous research: Theories, practices, and relationships*. Canadian Scholars Press.

Norgaard, K. M. (2019). *Salmon and Acorns Feed Our People*. Rutgers University Press.

Simpson, L. (2011). *Dancing on our turtle's back: Stories of Nishnaabeg re-creation, resurgence, and a new emergence*. Arbeiter Ring Publishing.

Truth and Reconciliation Commission of Canada. (2015). *Honouring the truth, reconciling for the future: Summary of the final report of the Truth and Reconciliation Commission of Canada*.

Tuck, E., & Yang, K. W. (2012). Decolonization is not a metaphor. *Decolonization, 1*(1), 1–40.

Waldram, J. B. (2008). The models and metaphors of healing: Introduction. *Medical Anthropology Quarterly, 22*(3), 243–245.

Chapter 7
Healing and Resilience:
Stories of Land–Based Connections

ABSTRACT

Chapter Seven positions Indigenous water healing as a multidimensional system of medicine grounded in ceremony, relational accountability, somatic transformation, and intergenerational knowledge. Through nine themes and fifteen subthemes, the chapter illustrates how water functions not only as a physical necessity but as a living relative, healer, and teacher. It fills major research gaps by centering ceremonial water practices, embodied healing, and emotional release—dimensions overlooked in biomedical, legal, and resource-focused water scholarship. The chapter highlights how grief, trauma, memory, and identity are carried and transformed through water-based rituals, while arts-based gatherings such as beading circles extend ceremonial healing into everyday life. It further emphasizes women's leadership, grandmother teachings, spirit naming, and the readiness required for water to receive emotional burdens.

INTRODUCTION

Building on the methodological foundations and governance frameworks established in Chapters 1 and 6, this chapter shifts from analytical explanation to lived experience. Rather than revisiting theoretical debates, it centers everyday ceremonial, emotional, and relational practices through which Indigenous healing and resilience are enacted in daily life. Existing work on environmental knowledge systems often focuses on ecological data or place-based narratives (Simpson, 2014; Kimmerer, 2013), but does not always link these to health and healing, particularly through the ceremonial acts of Elders, grandmothers, and water carriers. While some recent

DOI: 10.4018/979-8-3373-7559-5.ch007

scholarship has begun to explore water as a relational and gendered presence (Anderson, Clow, & Haworth-Brockman, 2013; Yazzie & Baldy, 2018), little attention has been paid to everyday ceremonial practices—such as water offerings, grief ceremonies, and spirit name retrieval—that restore not only individuals but entire communities. This chapter fills that gap by focusing on intergenerational transmission and emotional healing as enacted through water rituals. Moreover, while there is an emerging literature on Indigenous somatic and trauma healing (Duran, 2006; Hart, 2002), the intersection of water, trauma, and ceremonial embodiment remains underdeveloped. For example, water's role in processing stored grief, emotional regulation, and identity repair is seldom theorized within trauma-informed frameworks. This chapter contributes to the field by showing how water—through touch, immersion, tears, and offerings—supports somatic healing and emotional transformation. The discussion of grief, emotional readiness, and personal transformation through water builds upon recent work by Gone (2021) and Linklater (2014), who emphasize that healing must be culturally and spiritually anchored.

In terms of land-based learning, studies such as Wildcat et al. (2014) and Bang et al. (2016) have highlighted the educational and ontological dimensions of Indigenous relationships with the land. However, few studies explicitly address water-based ceremony as a site of teaching, identity formation, and governance. This chapter situates water not only as a site of healing but also as a teacher—a being that instructs, corrects, and witnesses. It elaborates how water healing affirms spiritual accountability and shapes ceremonial roles, such as the transformation into a kukum or water carrier. In doing so, this work complements and extends Indigenous land-based pedagogies into aquatic relations. Furthermore, Indigenous healing through water has rarely been addressed in cross-cultural and intercultural healing contexts. While some studies have examined Indigenous health interventions in urban or clinical settings (Smylie et al., 2009; Lavallee & Poole, 2010), there is a gap in literature discussing how water ceremonies facilitate relational healing across diverse communities. The examples presented in this chapter—including community gatherings, arts-based healing, and intercultural ceremonies—demonstrate how water becomes a unifying force that transcends boundaries and promotes collective transformation. These findings expand the scope of relational healing beyond the individual, toward a model of water-centered, community-based reconciliation.

Figure 1. How Cree woman knowledge-keeper explaining sustainable water and energy are inter connected with medicine for the community

Additionally, the commodification of Indigenous wellness practices is an increasing concern, particularly as land-based medicine becomes popularized in wellness industries. Scholars such as Tuck and Yang (2014) and TallBear (2019) have critiqued the extraction of Indigenous knowledge for capitalist gain. This chapter affirms that water healing must remain non-commercial, relational, and ethically accountable to ancestral law. By documenting teachings that reject profit-driven approaches to medicine and ceremony, it intervenes in ongoing debates around knowledge sovereignty and cultural appropriation. Through nine themes and fifteen subthemes, this chapter articulates a holistic, relational, and spiritually rooted understanding of water healing that directly responds to the gaps identified above. It moves beyond biomedical, legal, or extractive frameworks to affirm water as a relative and healer. It centers Indigenous women's knowledge, ceremonial protocols, and emotional readiness as vital to the healing process. It brings attention to the somatic and spiritual dimensions of water healing, offering an alternative paradigm for understanding trauma, identity, and community renewal. This chapter offers a necessary contribution to the growing field of Indigenous land- and water-based healing scholarship. While the governance and pedagogical dimensions of beading are discussed in Chapter 6, the focus here is on beading as an intimate, embodied practice through which emotional regulation, memory, and relational healing are sustained.

WATER, EMOTION, AND RELATIONAL HEALING

Ceremony, relational accountability, and water as relatives are defined in detail in Chapter 1. Here, these concepts are not re-theorized; instead, they are lived—through touch, memory, emotion, and presence in relation to water, land, and community.

Facilitating Healing Through Arts-Based Community Engagement

Arts-based community engagement provides a culturally grounded and spiritually significant pathway to healing. Beading circles, collaborative painting, and textile arts offer spaces where Indigenous people—especially women, Two-Spirit, and youth—gather to engage in shared acts of expression that are themselves ceremonial. These moments become sacred not because of ritual formalities but because of the shared emotional labour, intergenerational learning, and quiet reciprocity they involve. One Indigenous participant reflected, "It was like going through ceremony just to be in the process of making art and being with people." This quote illustrates how the atmosphere of artistic collaboration echoes the rhythms of ceremony: intentional, grounded, and spiritually open. Another participant noted, "Beading allows me to listen to my body—it's where my emotions settle." The physical movement of the hands, the silence that arises during creation, and the presence of others in the circle all facilitate emotional release and spiritual grounding. Beading becomes a medicine that restores balance, not through words, but through action, sensation, and collective care. This subtheme demonstrates that Indigenous arts are not separate from healing; they are an Indigenous methodology for processing trauma and restoring kinship.

INDIGENOUS HEALING THROUGH WATER

Settler colonial impact is not only Indigenous communities, but it has also been impacting on everything—it is a communal and spiritual journey that is often facilitated by water. In Indigenous worldviews, water is a living entity with the capacity to listen, hold, and transform emotional energy. As a sacred medium, water offers a space where grief can be acknowledged, expressed, and released. This theme explores the ceremonial and spiritual role of water in addressing emotional trauma and facilitating emotional clarity and balance.

Water as a Pathway to Healing Emotional and Physical Trauma

In many Indigenous traditions, water is not just a symbol of purification—it is an active participant in the healing process. Through acts like bathing in lakes, offering tobacco to rivers, or crying near sacred water sites, individuals connect their emotional states to the rhythms and listening capacity of water. An Elder shared, "I cried and cried by the water. It felt like the only place where my grief was allowed to move." This quote powerfully conveys the sanctuary that water provides for those navigating emotional pain. Water does not merely absorb tears; it transforms them.

Another speaker shared, "Our old people say water hears everything. It carries your pain away if you let it." In this worldview, grief is not something to be overcome, but something to be carried to water and ceremonially acknowledged. These teachings invite individuals to see water not only as physically restorative but as emotionally intelligent a being that can metabolize sorrow, frustration, and ancestral pain. Water ceremonies thus allow individuals and communities to restore flow—emotionally, spiritually, and relationally. This theme emphasizes that emotional healing is not simply a psychological process; it is a spiritual one, guided by relationship with water. Through ceremony, offering, and prayer, individuals surrender their emotional burdens to the water, participating in a cyclical, generational model of healing that affirms connection to land, spirit, and memory.

EMBODIED PRACTICES OF CARE

Healing in Indigenous traditions does not separate emotional, spiritual, and physical wellness. Hair and skin care are not cosmetic concerns—they are part of identity, ceremony, and wellness systems that have been handed down through generations. Water is central to these systems, used to activate the healing properties of plants, cool inflamed skin, cleanse the spirit, and nourish the body. This theme explores how water is integrated with land-based knowledge to treat hair, scalp, and skin issues, while simultaneously affirming the sacredness of the body.

Scalp Cleansing and Hair Growth

Hair holds memory, energy, and spiritual connection in many Indigenous cultures. Healing the scalp through water-based rituals is not just about physical rejuvenation—it is a way of releasing mental and spiritual clutter. An Indigenous participant shared, "She would boil herbs and pour it over our heads—it wasn't just about the hair, it was about healing the thoughts too." This reflects the holistic nature of Indigenous healing: the physical and metaphysical are intertwined. Herbs like sweetgrass, cedar, or sage—infused in water—are used to cleanse not only the scalp but also the mind, offering balance and clarity.

Moss and Skin Disorders

Moss, soft plants, and cool waters are traditional remedies for skin ailments such as rashes, inflammation, or chronic irritation. These are not simply herbal remedies—they are part of spiritual teachings. One Elder recalled, "My grandma would make a poultice with moss and wrap it with cloth—after two days the skin would cool and close." Moss, as a forest medicine, is valued for its gentleness and sacred humility. Its application is often done in ceremony, with prayer and intention. Here, water activates the healing compounds in plants while serving as a carrier of spirit. Water-based skin care affirms the body's dignity and supports mental and emotional well-being. It teaches that healing must begin with listening to the body and tending to its needs in slow, deliberate, and relational ways.

Internal and Topical Healing Combination

In Indigenous medicine, effective healing often involves treating the body from both inside and out. Water-based treatments such as teas, infusions, and baths are designed to work together holistically. One speaker explained, "You don't just wash the wound—you also drink the tea. That's how it knows where to go." This statement illustrates the principle that water is a spiritual messenger. When used internally and externally, water carries the intention of healing and communicates it to the body's energy systems. Healing the skin or scalp, therefore, is not simply about treating the visible—it is about restoring inner equilibrium. As Elders explained that drinking kidney-cleansing teas, applying salves, and praying during these practices complete the spiritual circuit of healing. Internal and external care are not divided; they are integrated as acts of self-respect and ancestral continuity.

Together, these three themes reveal the complexity and sacredness of Indigenous water healing. Whether through art, grief ceremonies, or embodied physical care, water becomes more than a natural element—it becomes a healer, relative, and teacher.

INDIGENOUS LAND-BASED SPIRITUAL IDENTITY AND MEDICINE WORK

Water is central not only to community healing but also to the development of individual spiritual identity and ceremonial responsibility. This theme addresses how personal encounters with water—through fasting, immersion, or prayer—lead to profound transformations, including the reception of spirit names and the assumption of medicine roles within community contexts.

Water in the Personal Healing Journey

Water is often described in Indigenous teachings as a sentient being, one that listens, responds, and guides. This subtheme elaborates on how water acts as both witness and healer in personal spiritual journeys. People turn to water not simply to cleanse but to be seen, affirmed, and restored in their identities.

As reflected in a powerful testimony: "That water knew my story before I spoke it. All I had to do was listen." This quote illustrates how healing through water involves deep listening, not only to one's own inner voice but to the spirit world mediated by water's energy. Water becomes both the mirror and the messenger, offering clarity, affirmation, and ancestral connection.

Receiving the Spirit Name

The act of receiving a spirit name is a sacred moment that often unfolds in proximity to water. This subtheme explores how ceremonies near rivers, lakes, or other sacred waters create openings for spiritual naming and identity revelation. These names are not assigned by human decision alone but are transmitted through dreams, visions, and natural signs during water-based fasts or ceremonies. A deeply moving quote from this chapter reveals, "It was during a fast by the lake that I received my name. Water carried it to me." Here, the lake functions not merely as a backdrop but as an active conduit of spiritual transmission. Naming becomes an act of reclamation—of spirit, lineage, and purpose—anchored through the presence of water as both witness and guide.

INDIGENOUS TRADITIONAL INNER HEALING AND WATER

Healing through water is not automatic; it requires readiness, intentionality, and emotional surrender. This theme discusses how water-based healing is contingent upon the internal state of the individual. Emotional preparedness and spiritual openness are essential for water to carry grief, tension, or illness away.

Self-Cleansing and Emotional Readiness

This subtheme emphasizes the reciprocal nature of healing—water can only take what a person is willing to release. Healing through water rituals begins with inner permission, where individuals consciously choose to surrender what no longer serves them. The chapter shares this reflection: "Water can't take what you still want to hold on to. You have to let it go first." This quote reinforces the agency and spiritual responsibility embedded in water-based healing. Water responds not to coercion but to authentic emotional offering. It requires participants to meet the ceremony with honesty and vulnerability, allowing the sacred act of release to unfold in its own time.

INDIGENOUS TRAUMA, MEMORY, AND SOMATIC HEALING

Trauma in Indigenous perspectives is not solely a psychological experience—it is also spiritual, physical, and somatic. This theme explores how water ceremonies facilitate the release of trauma that has become embedded in the body over time. These rituals do not merely soothe; they dislodge, transmute, and cleanse at the level of spirit and cellular memory.

Trauma Stored in the Body and Released Through Water Rituals

The subtheme describes how crying into water, soaking in sacred springs, or even silently holding a bowl of water can become pathways for releasing pain stored in the nervous system. Water acts gently but deeply, reaching emotional and somatic layers that verbal therapies may not touch.

As one speaker said, "Sometimes just holding the water in your hands is enough. It starts to draw out what your body's been keeping for years." This quote captures the subtle, patient, and deeply intuitive work that water does in ceremony. Water rituals support the body in becoming a site of restoration, where pain can finally exit and spirit can re-enter in calm and balance.

LAND-BASED ORIGINS AND PERSONAL JOURNEY WITH WATER CEREMONY

Water ceremonies are not learned solely through formal instruction but through family lineage, story, and witnessing. This theme highlights how ceremonial knowledge is first modeled by grandmothers, aunties, and Elders, then deepened through personal transformation and spiritual calling.

Family Teachings and Ceremonial Lineage

Indigenous ceremonial life is often passed down not through lecture, but through quiet daily acts. This subtheme underscores how watching a grandmother offer water to the sun each morning, or witnessing a mother boil cedar for spiritual baths, plants the seeds of ceremonial consciousness in the hearts of youth. "My grandmother never explained it, but I watched her every morning with the water. I knew it was important long before I asked why." This quote reveals that ceremonial knowledge is embodied before it is articulated. It speaks to the power of intergenerational witnessing in sustaining spiritual practices over time.

Personal Transformation and Becoming a Kukum

Ceremonial leadership is not claimed—it emerges through readiness, ancestral calling, and lived experience. This subtheme explores how the journey to becoming a kukum or ceremonial leader unfolds through healing crises, visioning, and the patient acceptance of one's spiritual role. "When I led my first ceremony, I realized I wasn't just doing what they taught me—I was becoming the one they taught." This reflection beautifully illustrates the sacred transition from student to carrier. Through water, ceremony, and spirit encounters, the learner becomes a lineage holder—an embodiment of memory and continuity.

This perspective critically illustrates that Indigenous water healing is not peripheral—it is central to restoring individual, collective, and spiritual wellness.

These sections reaffirm water's agency, its intelligence, and its relational capacity to hold grief, carry identity, foster ceremony, and release intergenerational trauma. Rooted in family teachings and extended through community engagement, water-based healing is both ancient and urgently contemporary—offering a paradigm of care, reciprocity, and transformation that resists the commodification of health and instead centers humility, reverence, and the sacred relational life.

INDIGENOUS LAND-BASED CONNECTION TO THE WATER AND HONORING GRANDMOTHER

This theme explores the sacred relationship between Indigenous peoples and water as an embodiment of grandmother spirit. Water is not just a physical substance but a spiritual presence that sustains life, memory, and ceremony. Within this worldview, water holds wisdom, emotional balance, and ancestral presence, and thus, daily interactions with water are grounded in gratitude, humility, and disciplined care. Rather than being treated as a resource, water is honored as a living relative—especially as a grandmother who watches over and nourishes both the body and the spirit. Through daily offerings and ceremonial attentiveness, Indigenous knowledge systems teach that water is a bridge between earthly life and the spiritual realm.

Daily Offerings and Personal Practice

Daily water offerings are a central expression of spiritual continuity, rooted in long-standing Indigenous traditions of relational accountability. These practices are not grand gestures; rather, they are small, disciplined acts that reaffirm gratitude and connection to the sacred each day. Offering tobacco, speaking to the water, or placing one's hands on a stream are examples of embodied reverence that help individuals stay spiritually grounded. These personal rituals sustain not only the person's connection to spirit but also reinforce relational ties to land, ancestors, and water beings. One community member shared, "Before I even brush my teeth, I speak to the water and give thanks. That's what keeps me grounded." This quote exemplifies how spiritual practice is embedded in the rhythm of daily life. Speaking to water before beginning one's morning routine signifies the prioritization of spiritual connection and ancestral respect. In this way, water ceremonies are not occasional but continual—alive in every gesture, every morning offering, and every act of mindfulness. These offerings do not require formal settings or elaborate protocols; their power comes from sincerity, intention, and regularity. They represent a worldview where every day begins in relationship—with land, spirit, and kin.

Respect and Everyday Ceremony

Respect for water is practiced not only in sacred gatherings but also through the most ordinary activities. How water is stored, poured, or discarded becomes a ceremony in itself. This subtheme reflects the belief that humility and spiritual attentiveness are carried through the handling of water in domestic life. Washing dishes, watering plants, or making tea are all actions that can be infused with respect and mindfulness. These everyday practices carry forward ceremonial teachings passed down by grandmothers, aunties, and Elders. As one Elder explained, "Even when washing dishes, I remember not to waste water—that's a kind of prayer too." This quote encapsulates how intention transforms everyday tasks into spiritual acts. By avoiding waste, acting with care, and remaining attentive to water's sacred role, individuals practice ceremonial living. These small acts of respect reaffirm the role of water as a teacher, a relative, and a spiritual force. Ultimately, this theme reminds us that water ceremony is not only found in ritual but lived in every moment we engage with the water.

LEARNING REFLECTION: RECLAIMING WATER AS CEREMONY, IDENTITY, AND RELATIONAL HEALING

This chapter offers an expansive re-centering of Indigenous water-based healing as a system of knowledge, ceremony, embodiment, and intergenerational care. Structured around nine themes and fifteen subthemes, it opens space for understanding water as a living entity with its own intelligence and ceremonial agency. The findings remind us that water is not merely symbolic nor resource-based—it is a relative, healer, witness, and teacher. Importantly, the learning offered here moves beyond surface-level engagement with water and enters deeply into Indigenous spiritual, physical, and emotional paradigms of healing. It is through this lens that water becomes ceremony, not in metaphor, but in lived practice.

One of the most reflective learnings from this chapter is that Indigenous water healing is not episodic or crisis-oriented—it is embedded in daily rhythms, familial relationships, personal readiness, and ceremonial responsibility. This stands in contrast to dominant medical models that often isolate treatment from spirit, or healing from land. Through the stories, teachings, and voices of Indigenous Elders, knowledge keepers, and water carriers, we come to understand that water-based healing is not a discrete act but a lifelong relationship. This relationship begins with humility and is sustained through daily offerings, quiet disciplines, and intentional presence.

Water, as shared by several speakers in the chapter, "knows your story before you speak it." This statement becomes foundational in understanding the epistemology of

Indigenous healing—it is not extracted or imposed, but relational and co-constructed. Water hears, responds, and reflects. Healing occurs not by force but by surrender. Whether it is crying by the lake, bathing in herbal infusions, or offering gratitude before drinking, these water interactions are less about treatment and more about relationship—about being in kinship with the natural world.

The chapter also revealed that water ceremonies are often the backdrop or setting for pivotal personal transformation. From receiving spirit names during fasting near lakes, to becoming kukum after years of water-based prayer, participants shared that healing is not simply something you receive; it is something you become. Water reveals identity, spiritual calling, and readiness. This is particularly evident in stories where individuals were not taught healing practices through explicit instruction but through witnessing their grandmothers' water rituals—boiling cedar, offering tobacco, or whispering to the river each morning. These acts seeded future ceremonial leadership. It is in these quiet, consistent, and often invisible forms of learning that Indigenous water knowledge is transmitted. It is intergenerational and somatic, residing in memory, gesture, and rhythm.

Another significant dimension of learning is the way this chapter framed art-based practice—especially beading—as ceremony. Beading, often led in collective circles, is not only a creative act but a healing one. Beading allows emotions to settle in the body. As one speaker shared, "Beading allows me to listen to my body—it's where my emotions settle." These tactile, repetitive movements are described as healing pathways that mirror the flow of water—steady, focused, rhythmic. When combined with the presence of others, such as in beading circles, this becomes a communal water practice—one of reciprocity, emotional processing, and spiritual alignment. This affirms the idea that water healing does not only occur by lakes or rivers, but in every act that mimics water's principles: flow, patience, and interconnection.

The relational foundation of water healing was further amplified in community gatherings and intercultural ceremonies. One of the most beautiful learnings here is that water does not discriminate. It speaks to everyone, holds everyone, and invites everyone into relation. Participants described how even non-Indigenous visitors to ceremony reported feeling an "energetic shift" simply by being in water's presence. Water ceremonies became bridges—reweaving connections among Elders, youth, families, and allies. This collective, inclusive aspect of healing demonstrates the transformative power of water to dissolve not only pain but also isolation, difference, and alienation. In this way, the chapter reframes healing as a decolonial, and cross-cultural act rooted in respect for Indigenous relational teachings.

Equally moving were the insights into grief and trauma held in the body. Participants described water as a site for releasing pain stored in the nervous system—grief that had lingered for years. One speaker noted, "Sometimes just holding the water in your hands is enough. It starts to draw out what your body's been keeping for years." This statement challenges the western notion of healing as verbal disclosure or therapeutic intervention. Instead, it reveals a somatic and spiritual approach where the body leads, water listens, and transformation occurs silently. These teachings remind us that trauma is not just mental—it is cellular. And water, with its gentleness, patience, and reciprocity, becomes a somatic healer.

In addition, the chapter warns against the commodification of healing—a theme echoed in warnings from Elders who reject the selling of medicine or ceremonial practices for profit. As one Elder said, "If you want to sell medicine, don't come to my camp." This quote is a powerful reminder that Indigenous water healing must remain accountable to spiritual law, community responsibility, and relational integrity. Healing is not a commodity; it is a ceremony. This perspective is crucial in a time when wellness industries continue to appropriate Indigenous knowledge without permission or responsibility.

The chapter also affirms that healing is not always dramatic. It can be gentle, slow, and ordinary. It can happen while washing dishes, whispering thanks to water, or rubbing a child's head with cedar tea. These everyday ceremonies are as sacred as any large-scale ritual. They reaffirm the foundational principle that healing happens through presence—being in intentional, relational practice with water every single day.

Finally, what this chapter teaches us—above all—is that water is not waiting to be understood in western terms. It is already speaking. The real question is whether we are listening. Listening, as emphasized throughout the chapter, is not passive. It is active, accountable, and reciprocal. Water listens to us; we must listen back. Only then does healing occur—not as a transaction, but as a return to kinship. In a time of ecological collapse, mental health crises, and cultural fragmentation, the teachings in this chapter feel urgent and generous. They offer a worldview where healing is not a response to dysfunction but a daily discipline of connection. Water, in this worldview, is not only what sustains life—it is what restores it. It teaches, holds, carries, and transforms. In return, all it asks is that we show up with care, with readiness, and with reverence. This chapter has not only documented water healing—it has enacted it. Through its teachings, stories, and ceremonial insights, it invites us to re-enter relation—with land, water, and ourselves. It reminds us that healing is not something we chase; it is something we remember. Water does not need us to fix it. It needs us to return.

REFERENCES

Anderson, K., Clow, B., & Haworth-Brockman, M. (2013). *Carrying water: Aboriginal women, water and health*. Atlantic Centre of Excellence for Women's Health.

Bang, M., Marin, A., Faber, L., & Suzukovich, E. (2016). Repatriating Indigenous land-based pedagogies. *Harvard Educational Review, 86*(1), 1–24.

Duran, E. (2006). *Healing the soul wound: Counseling with American Indians and other Native peoples*. Teachers College Press.

Gone, J. P. (2021). *A complementarity model for Indigenous healing*. In *APA handbook of multicultural psychology* (Vol. 2). American Psychological Association.

Hart, M. A. (2002). *Seeking Mino-Pimatisiwin: An Aboriginal approach to helping*. Fernwood Publishing.

Kimmerer, R. W. (2013). *Braiding sweetgrass: Indigenous wisdom, scientific knowledge, and the teachings of plants*. Milkweed Editions.

Lavallée, L., & Poole, J. (2010). Beyond recovery: Colonization, health and healing for Indigenous people in Canada. *International Journal of Mental Health and Addiction, 8*(2), 271–281. DOI: 10.1007/s11469-009-9239-8

Linklater, R. (2014). *Decolonizing trauma work: Indigenous stories and strategies*. Fernwood Publishing.

Simpson, L. (2014). Land as pedagogy: Nishnaabeg intelligence and rebellious transformation. *Decolonization, 3*(3), 1–25.

Smylie, J., Williams, L., & Cooper, N. (2009). Culture-based practices in Indigenous maternal and child health. *Canadian Journal of Public Health = Revue Canadienne de Santé Publique, 100*(3), 229–232.

TallBear. K. (2019). *Caretaking relations, not American dreamings*. In *Critically sovereign: Indigenous gender, sexuality, and feminist studies*. Duke University Press.

Tuck, E., & Yang, K. W. (2014). R-Words: Refusing research. In *Humanizing research: Decolonizing qualitative inquiry with youth and communities* (pp. 223–248). SAGE.

Wildcat, D., McDonald, J., Irlbacher-Fox, S., & Coulthard, G. (2014). Learning from the land: Indigenous land-based education. *Decolonization, 3*(3), 1–15.

Yazzie, E., & Baldy, C. R. (2018). Diné water politics and relational sovereignty. *Wicazo Sa Review, 33*(1), 6–28.

Chapter 8
Protecting Water as Responsibility for Moving Forward

ABSTRACT

This chapter advances a relational framework for understanding water as a living relative whose governance relies on Indigenous land-based science, ceremony, and reciprocal responsibility. It traces a thematic arc that moves from core principles of relationality to a forward-looking vision of Indigenous resurgence as sustainable water governance. Drawing on teachings from Elders, community-led practices, and influential Indigenous scholarship, the chapter reframes water governance as an ethical and ceremonial practice rather than a technical exercise. It emphasizes that climate change represents a profound relational rupture—disrupting languages, ecological relationships, and intergenerational knowledge systems. In response, the chapter argues for centering Indigenous sovereignty, water ethics, treaty responsibilities, and land-based education as essential pathways to sustainability. Policy recommendations for governments, universities, NGOs, and Indigenous nations highlight concrete mechanisms for restoring water justice and rebuilding community capacity.

INTRODUCTION

This book learning journey invites readers, students, and policymakers to walk through an Indigenous framework of understanding water, land, and climate as profoundly relational. It is not a journey of mastering concepts, but of entering relationships. For Indigenous nations across Western and Northern Canada, knowledge is not accumulated; it is lived. As the book demonstrates, learning is an ethical, ceremonial, and relational process that requires humility, accountability, and continuous renewal. The thematic arc below traces a conceptual movement from relational foundations to a vision of resurgence and governance that aligns with Indigenous water ethics and responsibilities.

As demonstrated throughout earlier chapters, Indigenous water governance is grounded in relational ethics and ceremonial responsibility. Relationality here does not function as an abstract concept, but as a lived system of accountability among water, land, people, and future generations. Ceremony operates as governance by guiding decision-making, regulating behaviour, and sustaining responsibilities to water as a living relative. Indigenous land-based science emerges from these relationships, producing knowledge through observation, care, and ethical engagement rather than extraction or control. This chapter does not re-theorize these foundations; instead, it turns toward how they can inform forward-looking policy and institutional transformation.

RELATIONALITY

As demonstrated throughout earlier chapters, Indigenous water governance is focused on relational ethics and ceremonial responsibility. Indigenous epistemologies begin with the understanding that the world is held together through relationships. Relationality is not a metaphor but a governance system, shaping how communities treat water, land, animals, and each other. The book illustrates that Indigenous Traditional Land-Based Knowledge is a relational science — one that emerges from thousands of years of living within ecosystems, observing seasonal rhythms, and maintaining ceremonial obligations to the more-than-human world. Relationality teaches that water is not an object but a living relative. Water is not a commodity but the movement of relationships within and between beings. Children learn these principles by being on the land, listening to Elders, and participating in water ceremonies that teach humility, gratitude, and reciprocity. When readers begin the book's journey through relationality, they enter an ethic of belonging. Knowledge becomes inseparable from responsibility; learning requires becoming accountable to the places that sustain life. This is the philosophical foundation that supports every subsequent theme — ceremony, governance, ethics, sovereignty, climate crisis, and resurgence.

CEREMONY

Across the book, ceremony emerges as one of the most powerful forms of Indigenous governance. Ceremony is not symbolic; it is a legal, scientific, ethical, and political system that shapes how communities maintain balanced relationships with land and water.

The learning journey leads readers to understand that ceremonies — such as water songs, spring offerings, fasting camps, seasonal feasts, and healing circles — hold encoded ecological knowledge. They document seasonal changes, water levels, migration patterns, and environmental disruptions. Ceremony also teaches restraint, humility, and responsibility.

Through the book's stories and frameworks, readers witness how water ceremonies led by women and grandmothers activate long-standing Indigenous laws that guide decision-making. Ceremony becomes a site of governance where teachings are enacted, power is shared, and obligations to water are renewed. For youth, participation in ceremony becomes a pathway into political and ecological leadership. Learning through ceremony requires readers to understand governance not as institutional process but as a relational practice grounded in spirit, memory, story, and land.

LAND-BASED GOVERNANCE

The next step in the learning journey encourages readers to rethink governance fundamentally. Indigenous land-based governance is not written in colonial statutes; it lives in practice, protocol, and story. Governance occurs when communities return to the land for teachings, harvesting, ceremonies, and collective decision-making.

Readers encounter examples such as:

- Cultural camps where Elders teach children how to offer tobacco, gather medicines, and speak to water.
- Community water walks that activate ancestral responsibilities.
- Language immersion environments where governance is spoken, not theorized.

These practices transmit Indigenous law in embodied form. Governance is not separated from daily life; it is lived through relationships of care and reciprocity.

Land-based governance also challenges technocratic approaches that dominate water policy. While colonial frameworks rely on extraction, measurement, and control, Indigenous governance emphasizes balance, restraint, and interdependence.

The book shows that sustainable futures require not only policy reform but a return to land-based governance as a foundation for environmental decision-making.

WATER ETHICS

The learning journey then dives deeper into Indigenous water ethics — one of the most important thematic pillars of the book. Indigenous water ethics do not conceptualize water as a resource but as a sacred being whose life sustains all others.

Readers learn that water ethics are expressed through:

- Language (terms for water embody law, story, and responsibility).
- Ceremony (water walks, offerings, songs).
- Environmental practice (ethical harvesting, monitoring, protection).
- Governance (decision-making rooted in kinship rather than private property).

Indigenous water ethics emphasize that water has agency, spirit, and rights. Humans are obligated to ensure its well-being. This ethic stands in stark contrast to extractive systems that treat water as property or a commodity for profit.

By entering Indigenous water ethics, readers encounter a worldview that teaches care rather than control. This ethical shift is essential for sustainable governance in a time of climate intensification.

SOVEREIGNTY AND TREATY RESPONSIBILITY

The thematic arc next leads readers into the political dimensions of Indigenous water governance. Sovereignty is not framed exclusively as jurisdiction or administrative authority — although these are important. Instead, sovereignty is taught as the collective responsibility to uphold relationships with land and water.

The book situates treaty responsibilities within this relational worldview. Treaties are not transactions but living agreements built on the principle of sharing, not surrender. Indigenous nations' responsibilities to water — and the responsibilities settlers assumed through treaty — remain active and binding.

Readers learn that honoring treaties requires:

- Returning authority to Indigenous nations over water governance.
- Recognizing that water sovereignty is inseparable from language, land, and ceremony.

- Confronting the failures of colonial governance systems such as pipelines, dams, mining, and boil-water advisories.

Understanding sovereignty in this relational way transforms environmental governance from a technical issue into a matter of justice, accountability, and ethical responsibility across generations.

CLIMATE CRISIS AS RELATIONAL RUPTURE

The climate crisis represents a profound rupture in Indigenous relational systems, disrupting responsibilities among water, land, animals, and people. As illustrated in earlier chapters, these disruptions are felt through ecological loss, emotional distress, and intergenerational disconnection. Rather than restating these impacts, this section emphasizes that climate crisis is not only an environmental condition but a governance failure—one that emerges when relational accountability is displaced by extractive and technocratic systems. Recognizing climate change as a relational rupture shifts the focus from documenting harm to addressing responsibility, creating the conditions for Indigenous-led governance responses articulated in the policy directions that follow.

INDIGENOUS RESURGENCE AS SUSTAINABLE GOVERNANCE

Indigenous resurgence offers a pathway beyond crisis by re-centering governance in Indigenous law, ceremony, and land-based responsibility. Across this book, resurgence is shown not as resistance alone, but as the active renewal of relationships that sustain water, community wellbeing, and future generations. Moving forward requires that sustainability efforts shift from managing impacts to restoring responsibility. Centering Indigenous land-based science within governance frameworks demands institutional transformation—recognizing Indigenous jurisdiction, supporting community-led monitoring, and embedding relational accountability into decision-making structures. This forward-looking framework does not call for inclusion within existing systems, but for the restructuring of governance itself so that Indigenous ways of knowing, governing, and caring for water shape policy, practice, and accountability.

MOVING FORWARD: CENTERING INDIGENOUS LAND-BASED SCIENCE AS SUSTAINABLE WATER GOVERNANCE

Table 1.

Actor	Core Responsibilities	Concrete Actions
Federal & Provincial Governments	Uphold Indigenous water sovereignty and treaty responsibility	• End all long-term boil-water advisories through sustained funding • Recognize Indigenous jurisdiction over watersheds • Legislate Indigenous water laws and rights of water • Implement FPIC as binding law for all water-related decisions
Indigenous Nations & Intertribal Alliances	Exercise self-determined water governance	• Establish Indigenous water councils led by women and Elders • Codify Indigenous water laws and protocols • Expand land-based water education and monitoring • Build inter-nation watershed governance networks
Universities & Research Institutions	Transform knowledge production and ethics	• Recognize Indigenous land-based science as scientific authority • Establish Indigenous-controlled ethics review processes • Fund long-term community-led water governance initiatives • Return university-held lands to Indigenous nations
Environmental NGOs & Organizations	Support Indigenous-led water protection	• Shift from advocacy *for* to advocacy *with* Indigenous nations • Redirect funding to Indigenous water protectors and guardians • Ensure all impact assessments are Indigenous led • End extractive research and symbolic inclusion practices

Moving forward into a future defined by deepening climate instability and intensifying water crises requires a profound transformation in how societies understand, relate to, and govern water. As this book argues, the current moment demands more than technical adaptation or managerial reforms; it requires a return to Indigenous land-based science as the foundation of sustainable water governance. Indigenous nations have shaped, protected, and governed water systems for thousands of years through relational ethics, ceremonial laws, and land-based pedagogies that centre responsibility, humility, and kinship. Re-centering these epistemologies is not a symbolic gesture but a political, ecological, and ethical necessity for collective survival (Borrows, 2020; Whyte, 2020). This section articulates a forward-looking framework for implementing Indigenous land-based science into water governance, drawing from ten influential studies in Indigenous water, climate, governance, and relational science. Together, these works illuminate how Indigenous knowledge systems offer sustainable governance models grounded in interdependence, reciprocity, and truth.

REFRAMING WATER GOVERNANCE THROUGH INDIGENOUS RELATIONAL SCIENCE

Indigenous land-based science rests on the foundational principle that water is a living relative — a being with spirit, agency, and legal standing within Indigenous legal orders. This is not metaphorical. As Kimmerer (2013) explains, Indigenous science views the natural world as composed of intelligent beings who teach humans how to live well. In this way, water governance emerges from relationships of care and reciprocity rather than control.

Whyte (2017) further expands this understanding by arguing that climate change for Indigenous peoples is not a future threat but an extension of ongoing colonial disruptions. Water injustice—seen through contamination, flooding, and dispossession—reveals the relational rupture between humans and the more-than-human world. Restoring Indigenous governance means restoring ethical relationships with water, grounded in kinship rather than resource management.

Similarly, Simpson (2014) emphasizes that Indigenous governance is enacted through land-based practices and ceremonial obligations. Ceremony functions as both scientific method and legal structure, guiding decisions about water protection and sustainable harvesting. These insights show that land-based science is not separate from governance; it *is* governance. Moving forward requires re-centering these relational frameworks at all levels of water governance — from community monitoring to provincial and federal legislation (LaDuke, 2020).

LAND-BASED SCIENCE AS PRACTICE: CEREMONY, MONITORING, AND ETHICAL RESPONSIBILITY

Indigenous land-based science is deeply empirical, rooted in long-term observation, seasonal knowledge, and ceremonial renewal. For instance, Arsenault et al. (2018) demonstrate how Indigenous communities conduct groundwater monitoring through embodied practices that implement observation, ceremony, and knowledge transmission. Such practices generate ecological data while reinforcing cultural responsibilities. McGregor (2021) similarly argues that Indigenous water governance must be understood as relational practice, where water protection work is guided by cultural identity, responsibilities to ancestors, and obligations to future generations. This differs fundamentally from Western science, which separates data from ethics, and measurement from meaning. Todd (2016) calls this separation a product of colonialism, arguing that water management must begin with Indigenous legal orders that recognize water as a rights-bearing entity. These legal orders are maintained through ceremony, story, and intergenerational knowledge transfer — all essential components of land-based science.

Thus, moving forward requires supporting Indigenous communities in revitalizing and expanding land-based monitoring systems that combine ceremony and science. These practices not only produce ecological data but sustain the ethical foundations of water governance.

THE ROLE OF INDIGENOUS LAW IN SUSTAINABLE WATER FUTURES

Indigenous law provides some of the most robust frameworks for sustainable water governance. Borrows (2020) describes Indigenous legal traditions as dynamic, relational, and grounded in land-based teachings. These laws emphasize reciprocity, respect, and responsibility — principles necessary for addressing escalating water crises.

Water governance cannot be truly sustainable without recognizing and implementing Indigenous legal orders. Simpson (2017) argues that sustainable futures require the resurgence of Indigenous governance systems, not merely their incorporation into colonial structures. Indigenous law is not a tool to supplement Western policy; it is a complete governance system that regulates relationships between humans and water.

INDIGENOUS WATER SOVEREIGNTY AS CLIMATE ADAPTATION

Climate change intensifies water insecurity, making Indigenous water sovereignty essential for adaptation. Whyte (2020) emphasizes that Indigenous climate leadership cannot be understood outside of their historical and ongoing experience of colonialism. Indigenous communities already have strategies for adapting to disruption — strategies grounded in community cohesion, land-based knowledge, relational governance, and cultural continuity.

In this context, water sovereignty becomes a framework for climate adaptation. It empowers Indigenous communities to:

- govern water based on cultural protocols
- protect watersheds and aquifers from extraction
- adapt to changing ecological conditions
- revitalize land-based education and monitoring
- restore ceremonial relationships disrupted by colonialism

When Indigenous governance leads, climate adaptation becomes a process of relational renewal rather than reactive crisis management.

Thus, centering Indigenous land-based science is not only ethical; it is necessary for climate survivability.

REVITALIZING INDIGENOUS LANGUAGE AS WATER GOVERNANCE

Language is a foundational pillar of Indigenous water governance. Kovach (2009) states that Indigenous methodologies — including water governance — emerge from language, story, and worldview. Without language, key teachings about water ethics, responsibilities, and protocols risk being lost.

Craft (2021) illustrates how Anishinaabe water law is embedded in language, ceremony, and relational concepts. Protecting water requires speaking about water in Indigenous languages that hold the stories, ethics, and relationships necessary for governance.

Moving forward requires supporting Indigenous immersion schools, land-based language camps, and childcare centres that teach children the relational vocabulary of water. When children regain the language of their ancestors, they also regain the governance systems that protect water.

REBUILDING COMMUNITY CAPACITY FOR LAND-BASED WATER GOVERNANCE

Indigenous water governance cannot be revitalized without community capacity, which has been weakened by colonial disruptions. As Latulippe and Klenk (2020) argue, Indigenous water governance requires structural supports, long-term funding, and political recognition — not short-term consultation processes. Furthermore, Anderson (2011, 2018) highlights the critical leadership role of Indigenous women in water protection and community governance. Moving forward requires empowering matriarchal leadership and supporting intergenerational teaching spaces where women and Two-Spirit knowledge keepers guide water governance processes.

To rebuild community capacity:

- Elders must be supported to teach land-based knowledge.
- Youth need access to land camps, water walks, and ceremonial learning.
- Communities require stable funding to develop Indigenous-led monitoring systems.
- Water governance boards must include Indigenous representation with decision-making authority.

Community capacity is not an add-on to governance: it *is* governance.

TRANSFORMING WESTERN GOVERNANCE SYSTEMS THROUGH INDIGENOUS PRINCIPLES

Western water governance systems must undergo structural transformation to align with Indigenous principles of relationality, reciprocity, and responsibility. Current systems prioritize economic growth, extraction, and technocratic management — practices incompatible with sustainable water futures. These pathways require courage, political will, and relational accountability. For sustainable water futures, Western systems must adapt to Indigenous law — not the other way around.

THE WAY FORWARD: A FRAMEWORK FOR CENTERING INDIGENOUS LAND-BASED SCIENCE

Drawing on the studies above, a forward-looking framework for sustainable water governance includes:

1. Centering Indigenous relational science

Water governance begins with recognizing water as a living being with rights and spirit.

2. Revitalizing ceremony as governance

Ceremony guides ecological understanding and ethical decision-making.

3. Supporting Indigenous water laws and sovereignty

   Recognizing Indigenous jurisdiction is essential for environmental protection.

4. Investing in Indigenous-led climate adaptation

   Communities must lead their own water governance practices.

5. Restoring language and cultural continuity

   Water governance is rooted in Indigenous linguistic and cultural frameworks.

6. Rebuilding community capacity

   Stable funding, land camps, and Elder-youth partnerships are essential.

7. Transforming colonial governance structures

   Governmental systems must center Indigenous law and relinquish unilateral control.

Moving forward requires more than inclusion — it requires transformation. Indigenous land-based science is not a supplement to Western governance; it is a world-sustaining knowledge system capable of guiding humanity through water crises and climate collapse. As Whyte (2020) argues, the future depends on relational accountability — to ancestors, to land, to water, and to future generations.

Indigenous water governance is Indigenous futurity.
It is sustainable governance.
It is climate adaptation.
It is justice.
This is the path toward relational water futures.

POLICY RECOMMENDATIONS FOR PROTECTING WATER, STRENGTHENING WATER SOVEREIGNTY, AND ADVANCING INDIGENOUS WATER JUSTICE

Despite decades of federal policy commitments, water insecurity remains a persistent and lived reality for many Indigenous communities in Canada. Recent federal reporting indicates that more than 32 First Nations communities continue to experience long-term boil-water advisories, affecting approximately 8,000 residents, many of whom have lived without consistent access to safe drinking water for years. Research demonstrates that these advisories are not the result of technical deficiencies alone, but rather stem from chronic underfunding, fragmented federal–provincial jurisdiction, and the systematic exclusion of Indigenous governance, law, and decision-making authority from water management processes (Phare, 2009). The persistence of boil-water advisories therefore reflects a structural failure of colonial water governance systems rather than isolated infrastructure breakdowns. This ongoing crisis underscores the urgency of centering Indigenous jurisdiction, land-based science, and relational accountability as foundational to water governance. Without addressing these governance inequities, technical interventions risk reproducing water injustice rather than achieving sustainable and equitable solutions (McGregor, 2021;).

Policy, in its most meaningful form, is not merely regulation, administration, or bureaucratic planning. Within Indigenous worldviews, policy is a living expression of relational responsibility — an enactment of the ethical, spiritual, and ecological obligations that bind humans, waters, lands, animals, ancestors, and future generations together. This chapter reframes policy as a relational tool: a pathway for restoring balance, repairing harm, and supporting Indigenous sovereignty in an era of escalating water crises and climate instability.

The recommendations here are not abstract guidelines; they are grounded in the teachings, narratives, ceremonies, and land-based practices at the heart of this book. They reflect what Indigenous communities, Elders, Knowledge Keepers, and land-based educators continually emphasize: sustainable water governance must emerge from Indigenous law, relational ethics, and community-led stewardship. Policies that fail to honour these foundations will continue to reproduce the colonial harms that have led to ecological degradation and widespread water insecurity.

I. Policy Recommendations for Federal and Provincial Governments

1. Recognize Indigenous Water Rights as Sovereign, Not Stakeholder-Based

Federal and provincial governments must formally recognize Indigenous water rights as inherent and sovereign. Indigenous nations are not stakeholders or consultees; they are original governors whose laws and practices have guided water stewardship for millennia.

Legislation must:

- Affirm Indigenous jurisdiction over waters within their territories.
- Recognize water as a rights-bearing entity with standing in law.
- Include Indigenous legal orders as equal to federal and provincial frameworks.

Without these legal recognitions, water governance will remain fragmented and unjust.

2. Implement "Free, Prior, and Informed Consent" (FPIC) as Binding Law

FPIC must be fully adopted and enforced in all water-related decisions. It must not be advisory or symbolic. Indigenous nations must possess:

- Veto power over major water infrastructure projects.
- Decision-making authority on watershed planning.
- Control over cumulative impact assessments.

FPIC must apply to pipelines, dams, mines, hydroelectric expansions, irrigation systems, and all industrial activity that affects water.

3. Restore Land and Watersheds to Indigenous Governance

Land return is not optional — it is central to water sovereignty. Returning control over watersheds, river systems, aquifers, and wetlands enables Indigenous nations to enact land-based governance rooted in relational ethics. Governments should:

- Identify priority watersheds for co-governance or full return.
- Support Indigenous-led watershed stewardship districts.
- Transfer federal and provincial lands to Indigenous nations.

This is the single most effective way to protect water long-term.

4. Legislate the Rights of Water

Following movements in Aotearoa, Ecuador, and portions of Turtle Island, Canada must legislate:

- The legal personhood of rivers, lakes, and aquifers.
- Water's right to flow, heal, rest, and be protected.
- Guardianship roles led by Indigenous nations.

This aligns with Indigenous teachings of water as a living relative.

5. Establish National Indigenous Water Protection Act

A federal Indigenous Water Protection Act should:

- Recognize Indigenous ceremonial law as governance.
- Prohibit harmful extraction in sacred waters.
- Create Indigenous water tribunals to resolve disputes.
- Fund Indigenous-led water protection programs.

Such an act must be co-written with Indigenous Knowledge Keepers and held accountable to treaty principles.

II. Policy Recommendations for Municipal and Regional Water Boards

1. Mandate Indigenous Representation with Decision-Making Authority

Municipal water boards often make decisions affecting Indigenous lands without Indigenous involvement. Boards must include:

- Indigenous Elders
- Water protectors
- Youth representatives
- Indigenous scientists and land-based educators

Representation must include decision-making authority, not symbolic advisory roles.

2. Center Indigenous Water Protocols into Local Water Systems

Municipalities should collaborate with Indigenous nations to implement:

- Water ceremonies
- Offering sites
- Seasonal protocols
- Indigenous naming practices
- Cultural monitoring practices

This fosters relational governance and supports cultural resurgence.

3. Ban or Restrict Industrial Activity within Sacred Waters

Municipalities must create bylaws that:

- Protect springs, wetlands, and ceremonial water sites.
- Limit waste discharge within culturally important waterways.
- Establish buffer zones around water bodies.

Such protections benefit the entire region.

4. Support Indigenous-Led Climate Adaptation Plans

Local governments should adopt Indigenous climate adaptation plans, which include:

- Land-based education for youth
- Watershed restoration
- Indigenous species protection
- Controlled burns and forest governance
- Community emergency planning rooted in relationality

Municipalities must fund and implement these plans fully.

III. Policy Recommendations for Universities and Research Institutions

1. Recognize Indigenous Land Based Knowledge as Scientific and Foundational

Universities must redefine "science" to include Indigenous land-based knowledge. This involves:

- Hiring Indigenous scientists, Elders, and ceremonial leaders.
- Offering land-based science degrees and programs.
- Building land-based learning camps on campus or community sites.

Such programs must be co-designed with Indigenous nations.

2. Establish Indigenous Sovereign Research Ethics Jurisdictions

Research involving Indigenous water, land, or knowledge must follow Indigenous ethics protocols.

Institutions must:

- Support Indigenous-controlled ethics boards.
- Require researchers to obtain community approval first.
- Ensure that knowledge shared in ceremonies is not extracted or misappropriated.

Ethics must protect relational knowledge as sacred.

3. Fund Long-Term Indigenous Water Governance Initiatives

Universities commit harmful extractive practices when they fund short-term projects that do not build community capacity.

They must instead:

- Provide stable funding for Indigenous water programs.
- Support Indigenous-led climate and water labs.
- Build partnerships that include land access, Elders' honoraria, and youth mentorship.

4. Return University-Controlled Lands to Indigenous Communities

This includes:

- Research stations
- Watershed parcels
- Forest tracts
- Sacred sites

Land return is essential to restoring Indigenous governance.

IV. Policy Recommendations for Environmental NGOs and Organizations

1. Shift from Advocacy "For" Indigenous Peoples to Advocacy "With" Indigenous Nations

NGOs must be accountable to Indigenous governance and knowledge. This means:

- Supporting Indigenous-led campaigns.
- Sharing funding and decision-making power.
- Ending appropriation of Indigenous symbols or teachings.

2. Redirect Resources to Indigenous Water Protectors

Funding must prioritize:

- Grandmother-led water walks
- Youth land-based education
- Community-based water monitoring
- Legal defense for water protectors

3. Ensure All Environmental Impact Assessments Are Indigenous-Led

EA processes must recognize:

- Indigenous law
- Cumulative impacts
- Cultural and ceremonial sites
- Indigenous environmental data

Only Indigenous nations can accurately assess the relational impacts of development.

V. Policy Recommendations for Indigenous Nations and Intertribal Alliances

1. Strengthen Indigenous Water Councils Rooted in Ceremony

Water governance councils should be:

- Led by women and grandmothers
- Guided by ceremony
- Grounded in language
- Connected across regions

Such councils form a foundation for nation-to-nation water policy.

2. Expand Land-Based Education as Governance

Land-based learning is governance. Nations should:

- Create immersive land camps for youth
- Revitalize water ceremonies
- Rebuild language fluency
- Support Elders to teach water ethics

This builds long-term sovereignty.

3. Strengthen Indigenous Watershed Governance Networks

Intertribal alliances can:

- Co-manage river systems
- Share data and protocols
- Build climate adaptation frameworks
- Advocate for water rights collectively

Waters flow across borders; governance must as well.

4. Draft Indigenous Water Laws and Codify Water Protocols

These should include:

- Community obligations to water
- Sacred site protections
- Intergenerational governance roles
- Water teachings and ceremonies
- Environmental responsibilities

These laws guide how nations govern water according to ancestral teachings.

VI. Cross-Cutting Recommendations for Water Justice and Sovereignty

1. Establish Indigenous Water Guardian Programs Across All Territories

Guardians maintain:

- Monitoring
- Education
- Law enforcement
- Stewardship
- Community water safety

These programs create jobs, cultural renewal, and ecological protection.

2. Fund Clean Water Infrastructure in All Indigenous Communities

Water justice requires eliminating:

- Boil-water advisories
- Infrastructure inequity
- Underfunded water systems

Colonial governments must provide sustained, not crisis-based, funding.

3. Recognize Water Justice as Climate Justice

Climate adaptation strategies must center:

- Indigenous land return
- Wetland and watershed restoration
- Food and medicine sovereignty

There is no climate justice without Indigenous sovereignty.

These recommendations call for more than institutional change — they call for a renewed covenant with water. Policy must be understood as ceremony, law, and relational responsibility. It must honour the living teachings held by Indigenous nations and repair the harms caused by colonial dispossession.

A sustainable water future requires:

- Indigenous sovereignty
- Land-based science
- Ceremonial governance
- Intergenerational learning
- Rights of water
- Climate adaptation rooted in relational ethics

This is not simply a political choice; it is a moral and ecological necessity. Protecting water is protecting life. Water sovereignty is the foundation of Indigenous resurgence. Water justice is the path to collective survival.

REFERENCES

Anderson, K. (2011). *Life stages and Native women: Memory, teachings, and story medicine*. University of Manitoba Press. DOI: 10.1515/9780887554056

Anderson, K. (2018). *A recognition of being: Reconstructing Native womanhood* (2nd ed.). Women's Press.

Arsenault, R., Diver, S., McGregor, D., Witham, A., & Bourassa, C. (2018). Shifting the framework of Canadian water governance through Indigenous research methods: Acknowledging water as a living entity. *Water (Basel)*, *10*(5), 452. DOI: 10.3390/w10010049

Borrows, J. (2020). *Law's Indigenous ethics*. University of Toronto Press.

Craft, A. (2021). *Anishinaabe Nibi Inaakonigewin Report: Reflecting the water laws and responsibilities of the Anishinaabe*. University of Manitoba Press.

Kimmerer, R. W. (2013). *Braiding sweetgrass: Indigenous wisdom, scientific knowledge, and the teachings of plants*. Milkweed Editions.

Kovach, M. (2009). *Indigenous methodologies: Characteristics, conversations, and contexts*. University of Toronto Press.

LaDuke, W. (2020). *To be a water protector: The rise of the Wiindigoo slayers*. Fernwood Publishing.

Latulippe, N., & Klenk, N. (2020). Making room and moving over: Knowledge co-production, Indigenous knowledge sovereignty, and the politics of global environmental change decision-making. *Current Opinion in Environmental Sustainability*, *42*, 7–14. DOI: 10.1016/j.cosust.2019.10.010

McGregor, D. (2021). Indigenous environmental justice and sustainability. In *Indigenous research: Theories, practices, and relationships* (pp. 76–94). Canadian Scholars.

Phare, M. A. S. (2009). *Denying the source: The crisis of First Nations water rights*. Rocky Mountain Books Ltd.

Simpson, L. B. (2014). Land as pedagogy: Nishnaabeg intelligence and rebellious transformation. *Decolonization*, *3*(3), 1–25.

Simpson, L. B. (2017). *As we have always done: Indigenous freedom through radical resistance*. University of Minnesota Press. DOI: 10.5749/j.ctt1pwt77c

Todd, Z. (2016). An Indigenous feminist's take on the ontological turn: 'Ontology' is just another word for colonialism. *Journal of Historical Sociology*, *29*(1), 4–22. DOI: 10.1111/johs.12124

Whyte, K. (2017). Indigenous climate justice and settler responsibility. *Daedalus*, *146*(3), 115–128.

Whyte, K. (2020). Too late for Indigenous climate justice: Ecological and relational tipping points. *Wiley Interdisciplinary Reviews: Climate Change*, *11*(1), e603. DOI: 10.1002/wcc.603

Chapter 9
Relational Water Governance in Practice:
As Sustainable Health Well Being

ABSTRACT

This chapter explores Indigenous water knowledge as a living system of relational governance in which relationships with water actively shape sustainability practices, community wellbeing, and collective responsibility. Moving beyond dominant Western frameworks that treat water as a resource or economic input, the chapter highlights how Indigenous governance understands water as a living relative, legal authority, and ethical guide whose health is inseparable from that of people and ecosystems. Grounded in principles such as "Water is life," Indigenous legal orders and stewardship practices position water at the center of identity, law, and ecological continuity. The chapter demonstrates how governance is enacted through ceremony, daily practice, oral teachings, and seasonal observation—operational mechanisms that regulate conduct, transmit ecological law, and sustain watershed health across generations.

INTRODUCTION

This chapter examines Indigenous water knowledge as a living system of relational governance, demonstrating how relationships with water actively organize sustainability practice, community health, and collective responsibility. Unlike dominant Western frameworks that conceptualize water primarily as a resource to be managed or an economic input to be optimized, Indigenous water governance understands water as a living relative, a legal and moral authority, and an ethical guide whose wellbeing is inseparable from that of people, land, and ecosystems (McGregor, 2021; Wilson & Inkster, 2018). Indigenous teachings commonly articulated through

DOI: 10.4018/979-8-3373-7559-5.ch009

principles such as "Water is life" reflect this relational ontology, positioning water as foundational to identity, law, and long-term ecological continuity rather than as a discrete environmental variable (Craft et al., 2019). Recent scholarship in Indigenous water governance highlights how Indigenous legal orders and stewardship practices emphasize interconnection among ecosystems, cultural lifeways, and governance responsibilities, challenging colonial governance structures that have historically marginalized Indigenous authority in water decision-making (Daigle, 2018; Simms et al., 2019). Rather than relying on centralized regulation or technocratic control, Indigenous governance systems are enacted through ceremony, daily water practices, oral teaching, seasonal observation, and collective stewardship. These practices function as operational mechanisms of governance that regulate human conduct, transmit ethical and ecological law, and sustain watershed health across generations (Wilson & Inkster, 2018; Yates et al., 2017). In contrast to extractive or control-oriented models, Indigenous governance prioritizes reciprocity, accountability, and shared care, fostering ecological resilience through long-term relational commitment.

Central to Indigenous water governance is the recognition that governance cannot be separated from lived practice. Indigenous communities across diverse territories continue to assert inherent rights and responsibilities to water through treaty relationships, traditional legal orders, and community-led monitoring systems that embed governance within everyday life (Craft et al., 2019; McGregor et al., 2020). In settler-colonial contexts such as Canada, these governance systems persist despite ongoing structural inequities, including long-standing drinking water advisories that reflect the legacy of colonial exclusion from statutory water decision-making (McGregor, 2021). Such conditions underscore the urgency of engaging Indigenous water governance models that center ethical obligations to water and collective wellbeing, rather than relying solely on crisis-driven or infrastructural responses.

Moving beyond descriptive or philosophical accounts, this chapter provides a clear translation layer for academic, policy, and practitioner audiences by demonstrating how Indigenous water knowledge functions in practice as environmental management and sustainability action. Indigenous relational approaches emphasize long-term ecological health, ethical limits, and shared obligations rather than short-term economic gain, offering sustainability pathways that are culturally grounded, ecologically responsive, and socially just (Whyte, 2020). This framing responds directly to calls within sustainability scholarship for governance models that integrate moral responsibility, intergenerational accountability, and ecological integrity into decision-making processes.

Water ceremonies, offerings, communal stewardship, and seasonal observation are examined in this chapter not as symbolic traditions but as practical governance mechanisms that establish responsibilities and ethical boundaries governing human interaction with waterscapes. Research demonstrates that Indigenous governance systems often succeed where colonial models falter because environmental protection is inseparable from community wellbeing and relational accountability (Simms et al., 2019; Wilson & Inkster, 2018). Through repeated practice, these governance mechanisms cultivate accountability at personal, communal, and intergenerational levels, ensuring that sustainability is enacted continuously rather than intermittently.

The chapter also addresses the interconnected physical, emotional, spiritual, and ecological dimensions of water-related wellbeing, including Indigenous responses to trauma, grief, illness, and climate-driven disruption. Indigenous communities interpret drought, flooding, seasonal shifts, and changes in water quality not merely as technical problems but as relational signals requiring adaptive, ethical responses grounded in ceremonial law and intergenerational responsibility (Whyte, 2020; McGregor et al., 2020). This perspective reframes climate change impacts as governance challenges rooted in disrupted relationships, rather than solely environmental anomalies requiring technical correction. Rather than treating sustainability as a technical intervention or emergency response, the chapter demonstrates that Indigenous water governance is continuous, relational, and enacted through everyday practices that integrate ecological observation, spiritual obligation, and community consensus. This approach aligns with emerging scholarship advocating for governance models that respect Indigenous knowledge sovereignty while engaging collaboratively with formal policy frameworks (Daigle, 2018; Yates et al., 2017). Importantly, Indigenous governance does not seek assimilation into Western regulatory systems but offers complementary pathways grounded in distinct ontological and ethical commitments.

Structurally, the chapter begins with community-based and ceremonial water practices and expands outward to collective, institutional, and policy-relevant dimensions of governance. This progression makes visible how Indigenous water knowledge translates into adaptable sustainability frameworks that can inform environmental planning, watershed stewardship, and legal reform without collapsing into program-specific prescriptions. Rather than competing with Western science, Indigenous water governance provides critical insights into relational ethics, ecological indicators, and long-term stewardship that can reshape sustainability practice when approached with humility and respect (McGregor, 2021). Finally, the chapter critically examines how Indigenous water governance challenges dominant colonial and capitalist sustainability models. By grounding governance in sacred law, humility, and relational accountability, Indigenous systems resist the commodification and technocratic management of water (Craft et al., 2019; Wilson & Inkster, 2018). Sustainability action is reframed as a moral and relational obligation—one that prioritizes ecological integrity, community wellbeing, and future generations over profit and short-term productivity. Drawing on teachings shared by Indigenous knowledge holders, this chapter demonstrates how water governance operates simultaneously at personal, communal, and institutional levels, positioning Indigenous water knowledge not as supplementary to sustainability discourse but as a leading framework for rethinking water sustainability in practice.

INDIGENOUS LAND-BASED MEDICINE AS HOLISTIC HEALTH AND HEALING

Indigenous land-based medicine is grounded in a holistic understanding of health that integrates body, mind, spirit, land, and relational responsibility. Rather than treating illness as an isolated biological malfunction, Indigenous healing systems understand wellness as the outcome of balanced relationships—among people, plants, ancestors, and place. Within this framework, medicine is not merely a therapeutic intervention but a **governance practice** that organizes care, responsibility, ethics, and sustainability across generations. This theme demonstrates how Indigenous land-based medicine functions as a comprehensive health system rooted in land, ceremony, and community knowledge. Through twelve interconnected subthemes, it illustrates how physical, emotional, spiritual, and relational wellbeing are sustained through daily practices of care rather than episodic treatment. These subthemes reveal that Indigenous medicine is inseparable from land stewardship, cultural continuity, and collective accountability, offering an applied model of holistic health that challenges biomedical, capitalist, and extractive approaches to healing.

Medicine as a Communal Right, not a Commodity

Indigenous medicine is fundamentally relational and communal; it is not owned, sold, or monetized. Healing knowledge is understood as a responsibility entrusted to individuals for the benefit of the people, not as intellectual property or marketable expertise. This principle directly challenges capitalist models that commodify care and transform medicine into a product. As one Indigenous knowledge holder states, *"The medicines are for the kids, for the people, for your loved ones. They are not for sale. So, I don't agree with that, and I absolutely refuse to work with anybody who is selling medicine for profit."* This teaching establishes medicine as a shared right and obligation, reinforcing governance through generosity, kinship, and accountability. Legitimacy in Indigenous medicine derives not from certification or profit, but from relational trust, ceremonial responsibility, and ethical conduct.

Blood and Immunity as Foundations of Healing

Indigenous healing systems understand blood as a carrier of ancestral memory, strength, and immunity. Health is therefore tied to lineage, ceremony, and inherited protection rather than solely to individual physiology. One Indigenous speaker explains, *"It's not so much about that your immune system is weak... the problem is always the blood. So, everything goes back to the blood."*

This teaching reflects an Indigenous epistemology in which immunity is relational and intergenerational. Blood holds the memory of survival, ceremony, and resilience, forming a foundation of protection that Western biomedical models often fail to recognize. Healing, in this context, involves restoring ancestral continuity as much as addressing present symptoms.

Emotional Well-Being as Integral to Physical Health

In Indigenous medicine, emotional balance is inseparable from physical healing. Medicines do not function independently of the emotional and relational state of the person receiving them. As one storyteller notes, *"Now we are living in a time right now, it's not just an effect of the environment. It can also be emotional."* This teaching affirms that grief, stress, fear, and relational rupture directly affect bodily health. Plant medicines are activated through emotional alignment, ceremonial readiness, and respectful relationships. Healing therefore requires attention to feelings, relationships, and context—not only symptoms.

Spirit-Led Healing and Ceremonial Calling

Healing roles within Indigenous communities are not self-appointed but spirit led. Knowledge holders are called through ceremony, dreams, and teachings rather than ambition or professional aspiration. During the COVID period, one speaker explains, *"That is what the spirit told us. Use stades and ratchets... if there is no bacteria in the body, they don't stay in the body."* This subtheme highlights that Indigenous medicine is guided by spiritual instruction that adapts to emerging conditions. Healing authority is accountable to spirit and community rather than institutional hierarchies, reinforcing a governance model based on responsibility rather than control.

Maternal Health Rooted in Ancestral Plant Knowledge

Indigenous maternal and reproductive health practices are deeply informed by land-based plant knowledge passed down through generations. One speaker explains the use of raspberry leaf, stating, *"Your womb actually looks like a raspberry... you need the raspberry leaves... before and after birth... it will prevent cancer in the womb."* This knowledge reflects precise anatomical and preventative understanding embedded within ceremonial practice. Maternal health is protected through sustained relationships with plants, ensuring continuity of life, lineage, and community wellbeing.

Organ-Specific Healing Through Traditional Knowledge

Contrary to assumptions that Indigenous medicine is generalized or symbolic, this subtheme demonstrates precise organ-based knowledge developed through observation and experience. One community member explains, *"Our stomachs are not supposed to be hard... now, because of all the flour in the intestine, we have people with rock-hard stomachs."* This teaching reveals detailed physiological understanding embedded in land-based medicine. Healing is targeted, specific, and responsive, guided by ancestral instruction and long-term observation of the body.

Restoring Internal Warmth Through Sun-Infused Medicine

Warmth is understood as a vital principle of wellness in Indigenous healing systems. Certain plants absorb and carry solar energy, transferring warmth back into the body. As one speaker explains, *"Maple buds really love the sun... they carry so much of the sun inside of them."* This teaching situates healing within seasonal, ecological, and energetic relationships. Medicine is not only biochemical but environmental and ceremonial, linking bodily regulation to land cycles.

Antibacterial Plants as Natural Protectors

Indigenous medicine includes sophisticated antimicrobial knowledge developed through long-term practice. One speaker explains the use of stades as a powerful antibacterial, noting that it *"strips both the good and bad bacteria in the body."* This knowledge demonstrates an understanding of microbial balance, risk, and application. Healing requires discernment and accountability, emphasizing that land-based medicine integrates both spiritual and biomedical efficacy.

Preventive Medicine as Everyday Practice

Healing in Indigenous communities is proactive rather than reactive. Medicine is taken regularly to maintain balance rather than waiting for illness. One speaker explains, *"You should be drinking the medicine, even if you are not sick."* This approach reframes health as maintenance, relationship, and foresight. Daily engagement with medicine sustains wellness and reduces crisis, offering a sustainability-oriented model of health governance.

Family-Centered Healing and Collective Care

Healing is communal, not individual. Medicine is prepared and shared among family members, including children and infants. As one speaker explains, *"I make it for my whole family... even the babies."* This subtheme emphasizes kinship-based care systems in which responsibility is shared, roles are flexible, and healing strengthens social bonds. Wellness is carried through family networks rather than isolated clinical encounters.

Listening to the Body as Diagnostic Practice

Indigenous diagnosis begins with listening—to the body, intuition, and relational cues. One speaker explains, *"I feel what my body needs... and I make the medicine."* This subtheme highlights embodied knowledge as a legitimate diagnostic tool, emphasizing attentiveness, self-awareness, and respect for bodily communication.

Oral Teachings Over Book-Based Learning

Finally, Indigenous medicine is learned through experience rather than textbooks. One speaker explains, *"All the plants have a twin... one is good and one is poisonous."*

This teaching underscores the risks of decontextualized learning and affirms that medicine requires mentorship, humility, and relational trust. Knowledge is safeguarded through lived practice rather than abstraction.

Together, these subthemes demonstrate that Indigenous land-based medicine is not focused on isolated symptoms but on restoring balance across physical, emotional, spiritual, and relational domains. Medicine is not something to be consumed or commodified, but something to be lived with, carried, and shared responsibly. Through plants, ceremony, intuition, and collective care, healing becomes a daily practice embedded in governance, sustainability, and cultural continuity. Indigenous land-based medicine thus offers a powerful model of holistic health grounded in relationship, responsibility, and renewal.

INDIGENOUS LAND-BASED MENTAL HEALTH AND CLIMATE CHANGE

Mental health challenges in Indigenous communities cannot be understood in isolation from land, climate change, colonial disruption, and intergenerational responsibility. Within Indigenous worldviews, mental wellness is not solely an internal or individual condition; it is deeply relational, shaped by connections to land, family, ceremony, and collective identity. When land is damaged, when families are displaced, and when traditional teachings are interrupted, emotional and psychological imbalance increases. This theme explores how mental distress emerges from ecological, social, and historical rupture—and how healing unfolds through reconnection to land-based lifeways, compassion, and cultural continuity. Indigenous land-based mental health frameworks challenge dominant biomedical models by situating distress within broader systems of governance, environment, and responsibility. Rather than pathologizing individuals, Indigenous approaches recognize that grief, anxiety, addiction, and depression often reflect broken relationships with land, ancestors, and community. The six subthemes that follow demonstrate how mental wellness is restored through land reconnection, identity renewal, intergenerational healing, compassion-based care, and resistance to capitalist pressures intensified by climate change.

Mental Distress Arising from Disconnection from Land

Disconnection from ancestral land produces profound emotional and spiritual consequences. Land is not only a physical place but a source of identity, belonging, and regulation. When people are removed from their territories—through displacement, environmental degradation, or colonial policy—mental health is directly affected. As one Indigenous speaker explains, *"Anytime you are disconnected from the land or any you can't be part of it, that affects your mental health... Anytime they are removed from their natural environment, you know, that's going to affect them."*

This teaching links addiction, anxiety, and depression to ecological and social displacement. Within Indigenous epistemologies, land is a regulating presence that supports emotional balance. Reconnection to land is therefore not symbolic but foundational to mental healing and cultural continuity.

Loss of Identity Through Colonization

Colonial systems fracture Indigenous identity by severing relationships to land, language, and ancestral knowledge. This subtheme highlights how imposed education, assimilation policies, and cultural suppression generate confusion, self-doubt, and psychological distress—particularly among youth. One speaker notes, *"Because of colonization, [people] are so disconnected from their land and our way of life that they've forgotten who they are, loss of identity."* Mental health challenges are thus inseparable from colonial disruption. Healing involves restoring identity through land-based teachings that replace external definitions with ancestral truths, reaffirming who people are and where they belong.

Intergenerational Trauma and Its Embodied Impact

Colonial violence does not end with one generation; it embeds itself in bodies, memories, and even biological inheritance. This subtheme addresses how trauma is transmitted emotionally, spiritually, and physically across generations. As one speaker explains, *"Genetically, we are being affected... trauma damages the brain and the heart... we are seeing the extreme of all of that in our new generations."*

This understanding aligns with emerging research on epigenetics while remaining grounded in Indigenous teachings that recognize trauma as embodied and relational. Healing must therefore work simultaneously through story, ceremony, and bodily reconnection to interrupt cycles of inherited pain.

Healing Through Reconnection with Traditional Practices

Reconnection to traditional lifeways restores emotional clarity, grounding, and belonging. Even small returns to land-based practices—such as medicine picking or ceremonial walks—can generate profound mental healing. One speaker share, *"When I got sober... going medicine picking... completely brought me back to myself. And that's something that no rehab... could ever do."*

This subtheme demonstrates that Indigenous healing does not rely on clinical intervention alone. Cultural reconnection functions as a primary form of mental health care, restoring coherence between identity, land, and spirit.

Compassion and Relational Understanding in Mental Health Care

Indigenous approaches to mental wellness prioritize compassion, listening, and relational inquiry rather than diagnosis and punishment. This subtheme contrasts Indigenous care practices with Western systems that often respond to distress through surveillance, policing, or institutionalization. One speaker explains why youth sought them instead of counselors: *"The counselors would hear these things... get worried and call the cops... instead... they were able to come to me, talk about their stuff, cry... get through all those emotions."* Healing begins with trust and understanding. Pain is not criminalized or pathologized but honored as meaningful expression. This relational model creates safe spaces where healing can unfold without fear.

Capitalism, Climate Pressure, and Youth Mental Health

Capitalist systems intensify mental distress by promoting individualism, extraction, and relentless productivity—values that conflict with Indigenous relational worldviews. Climate change further compounds these pressures by destabilizing land-based livelihoods and futures. As one speaker states plainly, *"The capitalism. Youths are the victims of capitalism."* Burnout, anxiety, and disconnection among Indigenous youth are not personal failures but structural outcomes. Land-based healing restores collective rhythms of life that resist capitalist speed, isolation, and commodification, offering alternative pathways grounded in responsibility rather than profit.

Together, these subthemes demonstrate that Indigenous mental health challenges are not merely individual conditions but manifestations of ecological, historical, and systemic disruption. Climate change, land loss, colonization, and capitalist values contribute to collective distress, while healing emerges through reconnection—with land, ceremony, identity, and community. Mental wellness, within Indigenous frameworks, is relational and restorative rather than isolating or corrective. When people return to traditional practices, when communities listen with compassion, and when land relationships are restored, healing becomes possible. This theme affirms that mental health is not achieved through separation from land and culture, but through their renewal.

INDIGENOUS LAND-BASED HEALING, INTERGENERATIONAL TRAUMA, AND RESILIENCE

Indigenous land-based healing is inseparable from the histories of colonization, dispossession, and cultural disruption that have shaped Indigenous lives across generations. Intergenerational trauma did not emerge in isolation; it was produced through deliberate colonial policies that targeted land relationships, ceremonial practices, language transmission, and spiritual authority. This theme explores how Indigenous Peoples are addressing the enduring impacts of residential schools and colonial violence through renewed participation in ceremony, land-based practice, and cultural resurgence. Healing, in this context, is not solely about recovery from harm—it is also an act of resistance, renewal, and governance that restores identity, memory, and collective strength.

Indigenous approaches to healing recognize that trauma is carried not only in individual bodies but in families, communities, and lands. The suppression of ceremony and spiritual practice disrupted Indigenous systems of regulation, care, and accountability, producing long-lasting psychological, emotional, and spiritual consequences. By restoring ceremony and land-based teachings, Indigenous communities are reactivating governance systems that sustain wellbeing across generations. The subthemes that follow examine both the damage caused by colonial institutions and the central role of ceremony in rebuilding identity, resilience, and relational balance.

Effects of Colonization and Residential Schools

Colonial systems were designed to dismantle Indigenous governance by attacking spiritual and cultural foundations. Residential schools, in particular, sought to sever children from their families, languages, ceremonies, and land-based knowledge. This subtheme examines how these institutions caused deep and enduring psychological harm—not only through physical and emotional abuse, but through enforced isolation, loneliness, and cultural erasure. As one Indigenous representative explains, *"And then the schools destroyed that. My parents suffered just as much as we did in school. It wasn't so much that we were a lot of kids who suffered abuse, but for us it was the loneliness that just about killed us."* This testimony highlights that trauma extended beyond individual experiences of abuse to encompass the systematic destruction of relational worlds. Parents and caregivers, themselves shaped by colonial disruption, carried unresolved grief and loss that affected subsequent generations. The removal of children from land and ceremony fractured systems of care that had historically supported mental, emotional, and spiritual wellbeing. Disconnection from ceremony was not incidental to colonial harm; it was a central mechanism through which Indigenous governance structures were weakened.

The effects of residential schools continue to manifest in contemporary mental health challenges, identity struggles, and social dislocation. Trauma is transmitted through memory, behaviour, and silence, shaping how individuals relate to themselves, others, and the land. Understanding these impacts is essential for recognizing why healing must extend beyond individual treatment to include cultural, ceremonial, and land-based restoration. Reclaiming ceremony is therefore not simply cultural revival—it is a necessary response to structural violence and an assertion of Indigenous authority over healing and governance.

The Role of Ceremony in Healing and Strengthening Identity

Ceremony plays a central role in restoring identity, belonging, and spiritual continuity after generations of suppression. This subtheme highlights how land-based ceremonies—such as sweat lodges, songs, prayer, and ritual practices—reconnect individuals to community, ancestry, and place. Through ceremonial participation, people reclaim their roles within Indigenous worlds and re-establish relationships that colonial systems sought to erase. As one Indigenous speaker explains, *"So, there are ceremonies that go along with asking for rain to replenish and make this land healthy. So, all the ceremonies that are life-giving are healing ceremonies."*

This teaching underscores that ceremony is not only personal or symbolic; it is ecological, relational, and governance oriented. Ceremonies that call for rain, renewal, or balance reaffirm reciprocal responsibilities between humans and the natural world. Healing is thus inseparable from land stewardship and environmental sustainability. By participating in ceremony, individuals reconnect to collective memory and cultural law, restoring coherence between identity, spirit, and land.

Ceremony also functions as a space where intergenerational knowledge is transmitted and renewed. Songs, language, and ritual practice carry teachings that ground individuals in who they are and where they come from. For those affected by colonial disruption, ceremony offers a pathway to reassemble fragmented identities and restore spiritual strength. Healing here is not passive recovery but active participation in cultural resurgence. Through ceremony, Indigenous Peoples reassert sovereignty over their bodies, spirits, and lands, rebuilding resilience that extends across generations.

Taken together, these subthemes demonstrate that healing from intergenerational trauma cannot occur without restoring ceremony and land-based practice. Residential schools and colonial violence attempted to dismantle Indigenous governance by targeting spiritual systems and cultural transmission. Ceremony remains the means through which Indigenous Peoples remember, resist, and regenerate. Healing, in this sense, is both personal and political—a process of reclaiming relational worlds and rebuilding collective strength.

INDIGENOUS LAND-BASED SPIRITUAL CLEANSING AND PROTECTION

In Indigenous knowledge systems, wellness includes not only healing from illness but protection from harm. Spiritual cleansing and protective practices address imbalances that arise from negative energy, spiritual interference, or relational disruption. This theme highlights how land-based medicine functions as a defensive as well as restorative practice, safeguarding individuals, families, and communities. Such practices affirm that not all illness or distress is physical in origin; some forms of harm must be addressed through spiritual means grounded in respectful relationships with the land. Protective medicines are part of Indigenous governance systems that regulate boundaries, restore balance, and maintain spiritual safety. These practices are carried out through ceremony, prayer, and precise protocols that emphasize humility and accountability. The subtheme below focuses on one such practice—the use of wild rose root as a form of spiritual cleansing and protection.

Wild Rose Root for Spiritual Defense and Cleansing

Wild rose root is used in Indigenous ceremonial practice to remove harmful energies and restore spiritual balance. This subtheme describes how the root is prepared and used in cleansing rituals to address "bad medicine," whether arising from illness, jealousy, or spiritual interference. One Indigenous knowledge holder explains the practice in detail: *"You boil it for about half an hour. Then you remove it, and you soak your feet in there. So as soon as this person is done soaking their feet for half an hour and praying, asking that medicine to help them. And as soon as they're done, they have to take it outside, and they say, I give this to Mother Earth to take care of."*

This ritual demonstrates how physical action, prayer, and ecological reciprocity are integrated within Indigenous healing practices. The act of soaking, praying, and returning the medicine to the earth reflects a governance ethic grounded in respect, consent, and reciprocity. Healing is not extracted from the land but temporarily borrowed and returned with gratitude. The practice affirms that protection and cleansing are achieved through relationship rather than domination. Wild rose root teachings also reinforce the understanding that spiritual wellbeing requires active maintenance. Just as physical health is sustained through ongoing care, spiritual protection depends on attentiveness to relational balance. This subtheme illustrates how land-based medicine functions as a protective infrastructure, guarding against unseen harms while reinforcing ethical responsibilities to land and spirit.

Together, these themes demonstrate that Indigenous land-based healing addresses both historical trauma and ongoing spiritual vulnerability through ceremony, protection, and relational accountability. Intergenerational trauma cannot be healed through individual treatment alone; it requires restoring the cultural, ceremonial, and ecological systems that sustain identity and resilience. Spiritual cleansing and protective practices further affirm that wellness includes defense against harm, not merely recovery from illness.

Returning to ceremony, land-based medicine, and ancestral teachings, Indigenous Peoples are not only healing—they are governing. These practices restore balance, reaffirm sovereignty, and sustain intergenerational strength in the face of colonial disruption and environmental change. Healing, in this framework, is an act of resurgence that reconnects people to land, memory, and responsibility, offering enduring pathways toward collective wellbeing and sustainable futures.

INDIGENOUS LAND-BASED ACCOUNTABILITY AS KNOWLEDGE

Within Indigenous knowledge systems, accountability is not an abstract ethical principle but a lived, land-based responsibility that governs how healing knowledge is received, carried, and shared. Indigenous land-based healing operates within a framework of relational accountability in which medicine, ceremony, and teaching are inseparable from spiritual guidance, humility, and community obligation. This theme examines how accountability functions as a form of Indigenous governance, regulating who may carry medicine, how knowledge is transmitted, and under what conditions healing remains effective and ethical.

Unlike Western credentialing systems, which often rely on institutional certification or self-identification, Indigenous healing authority is conferred through spiritual encounters and demonstrated readiness. Accountability is maintained not through profit, recognition, or formal status, but through ongoing relationship with spirit, land, and community. The subthemes that follow explore how individuals are called into healing roles and why Indigenous healing resists commodification as a matter of ethical survival.

Becoming a Teacher through Spirit-Led Encounters

In Indigenous land-based traditions, the authority to carry and teach medicine does not arise from personal ambition or intellectual mastery. Instead, it emerges through spirit-led encounters—visions, dreams, ceremonies, or unexpected events—that signal a person's readiness to assume responsibility. This subtheme examines how such encounters function as regulatory mechanisms, ensuring that healing knowledge is carried only by those who are emotionally, spiritually, and ethically prepared. One Indigenous knowledge holder describes this calling clearly: *"Your job is to carry the medicine for the people. Your job is to travel all over and give them back that medicine. And to reteach them."* This statement reflects a governance structure in which knowledge is entrusted, not claimed. The role of the healer is framed as service rather than status, guided by obligation to the people rather than individual advancement.

Spirit-led encounters operate as a form of accountability because they require humility, discernment, and relational confirmation. Not everyone who seeks knowledge is meant to carry it. This system protects medicine from misuse by ensuring that teaching authority is grounded in lived experience, ceremonial responsibility, and sustained relational conduct. In this way, Indigenous healing governance prioritizes readiness and ethical alignment over access or accumulation.

Healing Not for Profit

The commodification of Indigenous healing knowledge poses a profound threat to its integrity and efficacy. This subtheme critiques capitalist frameworks that transform medicine into a marketable product, severing it from ceremony, community, and accountability. Within Indigenous governance systems, healing is a spiritual obligation rather than a business opportunity, and profit-driven motives are understood to compromise the medicine itself. An Indigenous speaker articulates this boundary unequivocally: *"If you want to sell medicine, don't come to my camp. Do you want to make a profit from medicine? Don't come around me. I'm not going to teach you anything. You want to disrespect and dishonor the medicine and take more than you need? One day, medicine won't like you."* This teaching establishes clear ethical limits around knowledge transmission. Healing loses its power when it is separated from reciprocity, humility, and care. This resistance to commodification functions as a protective governance mechanism. By rejecting profit-based models, Indigenous healing systems safeguard medicine from exploitation, dilution, and institutional capture. Accountability here is not enforced through contracts or regulations, but through spiritual consequence and relational withdrawal. Knowledge is sustained only when carried in good relation.

Taken together, these subthemes demonstrate that Indigenous healing is governed by accountability structures that prioritize relationship over recognition and responsibility over revenue. Land-based healing affirms that ethical alignment—not visibility or certification—is the foundation of legitimate knowledge. This theme reinforces the book's broader argument that Indigenous governance systems already contain robust mechanisms for regulating practice, safeguarding sustainability, and maintaining collective wellbeing.

INDIGENOUS LAND-BASED HAIR AND SKIN HEALING

Indigenous land-based healing practices attend to the body in its full relational expression, including hair and skin as sites of memory, identity, and spiritual presence. This theme explores how traditional plant-based care for hair and skin operates as part of holistic health systems that integrate physical treatment with emotional, ceremonial, and relational significance. These practices are not cosmetic interventions; they are governance practices that restore dignity, balance, and confidence while reaffirming relationships between body, land, and spirit. Hair and skin carry cultural meaning and energetic memory in many Indigenous traditions. Their care reflects how individuals relate to themselves and to the world. The subthemes below examine scalp cleansing and hair growth, the treatment of skin conditions using moss and land plants, and the combined use of internal and topical medicines as a unified healing system.

Scalp Cleansing and Hair Growth

In Indigenous knowledge systems, hair is understood as a carrier of memory and energy, and its care is considered spiritual work. This subtheme explores how traditional plant-based practices restore scalp health and encourage hair growth, often in contexts where hair loss is connected to stress, trauma, or disrupted identity. The process is not merely biological but ceremonial, supporting both physical regeneration and emotional renewal.

An Indigenous community representative describes the practice: *"So you get that, and you spray it on your head, and you rub your head upside down like this, and you keep massaging your head. What it does is it clears out the follicles in the hair and it unclogs them."* This explanation reflects detailed anatomical knowledge embedded within land-based healing systems. Scalp cleansing is understood as a process of clearing blockage—physically and energetically—allowing renewal to occur. Caring for the scalp becomes a ritual of restoration, helping individuals reclaim strength and presence after periods of illness, stress, or loss. Hair growth is thus linked to broader processes of healing and self-reclamation, reinforcing the relational nature of Indigenous health practices.

Moss and Skin Disorders

Mosses and other soft land plants are used in Indigenous medicine to treat a range of skin conditions, drawing on their cooling, soothing, and antimicrobial properties. This subtheme emphasizes that such practices are guided not only by observation but by spiritual instruction. Knowledge about when and how to gather and apply moss is transmitted orally and ceremonially, ensuring respectful engagement with the land.

One Indigenous speaker recounts a spirit-led instruction: *"And the spirits told me, go to the mountains and go get moss... take him to the water in the mountain and wash him up... and we wrapped him up."* This narrative illustrates how land-based healing responds to both surface-level conditions and deeper spiritual imbalance. Moss medicine addresses irritation, inflammation, and harm while restoring calm and protection.

These practices reflect an understanding that skin conditions may be expressions of relational disturbance, requiring care that attends to both body and spirit. Healing emerges through respectful collaboration with land beings rather than extraction.

Internal and Topical Healing Combination

Indigenous healing approaches recognize the body as an integrated system in which internal and external treatments must work together. This subtheme focuses on the combined use of internal medicines—such as teas and blood cleaners—and topical applications to support full-body healing. Treating symptoms without addressing internal imbalance is considered incomplete. An Indigenous caregiver explains: *"So on my grandson, we also did the blood cleaner for him... the kidney, the liver, and the blood cleaner... and we put the cream on every day."* This teaching demonstrates disciplined, sustained engagement with healing practices. Internal cleansing supports organ function and circulation, while topical care addresses surface conditions, completing the healing cycle. This layered approach reflects Indigenous governance principles that emphasize balance, consistency, and accountability. Healing requires commitment, patience, and relational care, reinforcing the idea that wellness is maintained through daily practice rather than episodic intervention.

Together, these subthemes demonstrate that hair and skin care in Indigenous land-based healing systems are deeply relational, ceremonial, and intergenerational. Restoring physical health on the surface is inseparable from restoring dignity, confidence, and spiritual clarity. Through plant knowledge, ritual practice, and attentive caregiving, bodies are cared for as living expressions of relationship with land and community. This theme reinforces the book's broader argument that Indigenous healing practices function as embodied governance systems. They regulate care, transmit knowledge, and sustain wellbeing through ethical, relational engagement with the land. In doing so, they offer enduring models of sustainability that honor the full humanity of those they serve.

LAND-BASED TRADITIONAL MEDICINES FOR PAIN AND INFLAMMATION

Indigenous land-based medicine includes highly specific, place-based approaches to managing physical pain and inflammation. These practices are grounded in intimate knowledge of plants, seasonal cycles, and bodily systems, and are transmitted through generations as part of everyday governance of health and wellbeing. This theme examines how Indigenous Peoples address chronic pain not through abstraction or generalized treatment, but through targeted remedies embedded in relational accountability, ceremonial care, and environmental knowledge.

Pain management within Indigenous healing systems is not isolated from ethics or responsibility. Medicines are gathered with intention, applied with discernment, and governed by protocols that protect both the individual and the integrity of the medicine. Birch, a widely present tree across many Indigenous territories, offers a clear example of how land-based knowledge functions as applied healthcare governance—regulating dosage, placement, and use through teachings that integrate anatomy, spirituality, and respect for the body.

Birch Paper for Arthritis

Birch bark–based paper has long been used in Indigenous communities to treat joint pain and inflammation, particularly in aging bodies. This subtheme focuses on how birch medicine exemplifies precision in Indigenous pain management, demonstrating that land-based healing is neither vague nor symbolic, but clinically attentive and ethically regulated. An Indigenous knowledge holder explains the application clearly: *"So then you take it out. Somebody has arthritis in their spine. They would put that on their spine before they go to bed. You can put it on the spine, but it can't be put over the heart in the heart area. It should never be placed there."* This instruction reflects detailed anatomical awareness and an understanding of bodily boundaries. The medicine is effective precisely because it is applied with care, discernment, and respect for vital organs. This teaching illustrates that Indigenous pain treatment operates within an internal logic of safety and responsibility. The prohibition against placing birch paper over the heart is not arbitrary; it reflects a holistic understanding of bodily vulnerability and energetic balance. In this way, Indigenous medicine governs practice through oral law rather than written prescription, ensuring that care remains relational and accountable.

Birch medicine also highlights the sustainability dimension of Indigenous healing systems. Birch is not harvested indiscriminately; its use is governed by seasonal awareness, gratitude, and restraint. Healing is inseparable from environmental stewardship, reinforcing the book's broader argument that Indigenous governance systems integrate health, land care, and ethical responsibility.

This theme reminds readers that Indigenous knowledge addresses everyday physical suffering through relational medicine that is both practical and deeply contextual. Birch, a familiar presence on the land, becomes a powerful healer when gathered with intention, applied with love, and remembered across generations. Pain relief here is not only about symptom management but about maintaining respectful relationships with the land that sustains life.

INDIGENOUS WAYS OF HEALING THROUGH MEDICINE AND CEREMONY

Within Indigenous knowledge systems, healing is governed by sacred law. The use of land-based medicine is never random, improvised, or detached from responsibility; it follows ceremonial protocols rooted in respect, reciprocity, and relational accountability. This theme emphasizes that Indigenous healing practices are structured systems of governance shaped by generations of teachings. Whether harvesting, preparing, or sharing medicine, each action is guided by ceremonial law that regulates human behavior in relation to land, spirit, and community. Indigenous medicine operates within a legal and ethical framework transmitted through oral tradition, lived practice, and ceremony. Healing is therefore not simply about efficacy or outcome, but about *how* medicine is engaged, *when* it is gathered, and *why* it is shared. These protocols ensure that healing remains balanced, ethical, and spiritually aligned, preventing extraction, misuse, or commodification.

Gathering Protocols and Sacred Law

This subtheme illustrates how sacred law governs medicine gathering. Healing practices must follow the laws of land, spirit, and ceremony—not personal convenience, urgency, or market demand. Knowledge of timing, technique, and offering is as essential as knowledge of the plant itself. As one Indigenous knowledge holder explains: *"Wait till the seed pods are open. So, when we pick and gather any medicine, we shake it. We wait for the seed pod to crack open, and we shake it upside down."* This teaching affirms that land-based healing follows precise protocols passed down from ancestors. The act of waiting is not passive; it is an expression of respect for plant agency, seasonal rhythm, and ecological regeneration. Following sacred law ensures that healing work remains relational rather than extractive. Medicine is not taken—it is received through proper relationship. In this way, ceremonial protocols function as environmental governance, regulating human interaction with land to sustain balance across generations.

INDIGENOUS LAND-BASED STRUGGLES AND LESSONS

The land holds stories of both injury and instruction. Healing does not require forgetting pain; it requires learning how to live, grow, and care responsibly in its presence. This theme explores how damaged landscapes—marked by colonial extraction, displacement, and ecological violence—remain sites of teaching, resilience, and renewal.

Indigenous healing acknowledges that land, like people, carries scars. These scars do not negate sacredness; they deepen responsibility. Healing emerges not from idealized notions of pristine nature, but from sustained relationship with land as it exists—wounded yet enduring.

Healing with the Scarred Land

It reflects on the relational parallel between human trauma and environmental damage. As one speaker shares: *"But when we spent our first winter out in Bay…*

we struggled... what we realized was that the land was just as scarred as we are." This realization reframes hardship as shared experience rather than failure. Scarred land teaches humility, patience, and perseverance. Returning to damaged places becomes an act of commitment rather than nostalgia. Healing includes grieving the land, witnessing its pain, and choosing to remain in relationship anyway. In this sense, land-based healing is not about restoration to an imagined past, but about learning how to live ethically within altered realities.

Indigenous Land-Based Trauma Causing Healing Interruption

Trauma disrupts presence, ceremony, and relational connection. It interrupts the flow of healing within and across generations, affecting not only individuals but families and communities. This theme examines how trauma manifests as absence—emotional, spiritual, and relational—even when people are physically present. Indigenous frameworks understand trauma not only as an internal psychological state, but as a relational rupture that blocks participation in parenting, ceremony, and caregiving. Healing therefore requires more than action; it requires restoring presence.

Trauma Blocking Presence and Parent–Child Connection

This subtheme focuses on how unresolved trauma interferes with emotional availability, particularly in parent–child relationships. One Indigenous speaker states plainly: *"That's what trauma does to you. It steals you away from your children and your grandchildren."* Trauma pulls individuals inward, creating numbness and distance even in moments of closeness. This teaching highlights that healing is not only about doing—attending ceremonies, gathering medicines—but about *being present*. Land-based practices, ceremony, and relational reconnection serve as pathways for returning to oneself, which is a prerequisite for reconnecting with others. Healing begins when presence is restored, allowing care, teaching, and love to flow again across generations.

LAND-BASED TRADITIONAL COMMUNITY HEALING AND SIGNS FROM NATURE

Indigenous healing systems are deeply responsive to the messages carried by the natural world. Animals, weather, water, and seasonal changes communicate guidance for healing practices and community decision-making. These signs are not metaphorical; they are ceremonial instructions embedded in ecological relationship.

This theme emphasizes that land-based healing requires attentiveness and humility. The Earth is an active participant in governance, offering warnings, confirmations, and timing through observable phenomena.

Environmental Change and Spiritual Signs

Elders and healers interpret environmental signs to determine when to act, gather medicine, or conduct ceremony. As one community member explains: *"It was a crane... for me I felt that it was a sign that our men are going to be starting to heal."* This interpretation reflects a relational intelligence developed through long-term engagement with land. Reading signs from nature is a spiritual obligation grounded in responsibility rather than control. These signs guide collective action and reinforce the understanding that healing unfolds in rhythm with the land's intention. Observing, waiting, and responding appropriately are forms of governance that align human behavior with ecological balance.

Together, these themes demonstrate that Indigenous land-based healing is governed by sacred law, relational accountability, and deep ecological intelligence. Healing follows protocols, responds to trauma, learns from damaged landscapes, and listens attentively to the natural world. These practices are not symbolic or peripheral; they constitute applied systems of governance that regulate behavior, sustain relationships, and support long-term resilience.

By centering ceremony, presence, and responsibility, Indigenous healing frameworks offer vital insights for sustainability, environmental ethics, and collective wellbeing. Healing, in this context, is not about perfection or erasure of pain—it is about learning to live well, listen deeply, and act responsibly within a living, relational world.

LAND-BASED TRADITIONAL HEALING FOODS AND NUTRITIONAL MEDICINE

Within Indigenous knowledge systems, food is medicine. Traditional land-based diets are not understood simply as sources of calories or nutrients, but as relational gifts that carry memory, emotion, ceremony, and responsibility. This theme centers the role of land-based foods in healing, emphasizing how nourishment supports long-term physical, emotional, and spiritual well-being. Food, in this context, is not fuel consumed in isolation; it is a reciprocal relationship between people, land, and ancestors that sustains life across generations.

Indigenous nutritional medicine reflects a holistic understanding of the body as inseparable from emotion, spirit, and environment. Foods are selected not only for their physical properties, but for their energetic impact, seasonal appropriateness, and ceremonial role. When gathered, prepared, and shared in accordance with sacred law, food becomes a form of governance—regulating care, reinforcing relational accountability, and sustaining community resilience.

Clover as Sweet Nutritional Support

This subtheme highlights clover as a gentle, nourishing medicine that supports both the physical and emotional body. Clover is often introduced to children, Elders, or those recovering from exhaustion, grief, or depletion. Its sweetness is not incidental; it reflects the plant's relational role as a comforter and stabilizer within Indigenous healing systems. As one Indigenous knowledge holder explains: *"So they used to pick them long time ago. And they'll dry all the clover. It tastes like honey, They're sweet. And especially when they're all ripe. they taste so good."* This teaching reveals how plants are understood relationally, with personalities and intentions that guide their use. Clover is not aggressive or corrective medicine; it offers reassurance, warmth, and nourishment at moments when the body and spirit need gentleness rather than intensity. Food-based healing, as demonstrated through clover medicine, reminds us that wellness does not always require intervention or treatment. Sometimes healing emerges through sweetness, care, and patience. When food is gathered respectfully and shared in ceremony, even the most common plants become powerful reminders of the land's generosity and the importance of thoughtful, balanced care. In this way, nutritional medicine functions as both sustenance and ceremony, reinforcing the principle that healing begins with how we feed ourselves and one another.

INDIGENOUS LAND-BASED PRACTICES AS MEN'S HEALTH AND PROSTATE HEALING

Indigenous land-based medicine addresses men's health through spiritually grounded, relational practices that support balance, longevity, and accountability. This theme emphasizes that men's healing is not limited to physical treatment; it is deeply connected to emotional responsibility, ceremony, and reconnection with land. Traditional knowledge includes specific remedies for conditions such as prostate inflammation and cancer, demonstrating the precision and depth of Indigenous plant medicine. Men's health, within Indigenous frameworks, is not treated in isolation from community or spirit. Healing requires men to return to ceremony, to relationship, and to accountability for their bodies and roles within the collective. Plant-based remedies function most effectively when embedded within these relational contexts.

Rabbit Root and Sage for Prostate Cancer

This subtheme focuses on the combined use of rabbit root and sage in addressing prostate inflammation and cancer. The pairing of these medicines reflects Indigenous approaches to layered healing, where multiple plants work together to support balance rather than target symptoms in isolation. As one Indigenous practitioner explains: *"It looks like an umbrella, but it's also good for prostate cancer, yes, and you would add it with sage, and that would heal prostate cancer."* This statement affirms the practical efficacy of traditional plant medicine while emphasizing the importance of relational caregiving. Healing is not performed alone; it unfolds through guidance, support, and accountability within ceremonial contexts. Men's healing, in this framework, includes emotional honesty, spiritual responsibility, and reconnection to land. Prostate care becomes an entry point for broader wellness, encouraging men to re-anchor themselves in ceremony and community. Through plant medicine and relational support, healing becomes a pathway back to balance rather than a response to crisis alone.

LEARNING REFLECTION: RELATIONAL HEALING AS A SYSTEM OF CARE

This chapter affirms that Indigenous land-based healing is not a marginal, alternative, or supplemental practice, but a comprehensive system of care and governance encompassing emotional, physical, spiritual, ecological, and political dimensions of life. Across fifteen thematic areas and thirty-six subthemes, the teachings presented here reveal a worldview in which land is not a passive setting for healing, but an active agent—functioning simultaneously as healer, Elder, teacher, and ceremony. Plants, roots, mosses, foods, stories, and songs are not resources to be extracted or techniques to be deployed; they are living relatives that carry memory, instruction, and ethical responsibility (McGregor, 2021; Wilson & Inkster, 2018).

One of the most significant insights emerging from this chapter is that healing within Indigenous knowledge systems is not always oriented toward curing in a biomedical sense. Rather, healing is often understood as a process of return—returning to the body, to memory, to relational responsibility, and to land. This process is cyclical rather than linear, relational rather than individualistic, and continuous rather than episodic (Whyte, 2020). Healing cannot exist in isolation from relationships with family, community, ancestors, and the more-than-human world. Whether enacted through beading, harvesting, bathing, listening, offering, or ceremony, healing is always a dialogue with place and an enactment of governance grounded in lived responsibility.

The teachings throughout this chapter directly challenge dominant Western medical and sustainability models that separate treatment from spirit, individualize responsibility, and commodify care. Indigenous knowledge systems assert that the legitimacy of medicine does not derive from institutional certification, market value, or professional status, but from relational accountability to land, ceremony, and community (Craft et al., 2019; Smith, 2020). Elders consistently emphasize that when healing becomes ego-driven, profit-oriented, or detached from ceremonial law, it loses its efficacy. Medicine, to remain medicine, must be practiced within ethical limits defined by land-based law and collective responsibility.

The chapter's engagement with trauma—particularly the impacts of residential schools, parent–child disconnection, and intergenerational harm—demonstrates that ceremony is not optional within Indigenous healing systems. Ceremony provides structure for grief, pathways for identity repair, and tools for spiritual reconnection that cannot be replaced by clinical intervention alone (McGregor et al., 2020). Importantly, ceremony is not confined to formal or institutionalized rituals. It may take the form of beading, silence, shared listening, food preparation, fasting, prayer, or presence without judgment. These practices reaffirm bodily and spiritual sovereignty, reminding individuals and communities that healing begins when the body is recognized as sacred once again.

Many of the subthemes also emphasize that Indigenous medicine is deeply sensory and embodied. Smell, texture, warmth, taste, and rhythm function as teachers, guiding healing through experiential knowledge rather than abstraction. Healing unfolds slowly, seasonally, and patiently, reinforcing temporal rhythms that resist the urgency, acceleration, and extraction characteristic of colonial and capitalist systems (Daigle, 2018). In this way, land-based healing reorients sustainability away from crisis management toward long-term relational care.

Gendered teachings further illustrate the complexity and nuance of Indigenous healing systems. Responsibilities related to women's health, men's health, grandparent teachings, youth learning, and water stewardship are not rigid roles imposed through hierarchy, but sacred responsibilities earned through teaching, spiritual encounters, and accountability (Simms et al., 2019). These roles reinforce balance rather than dominance, emphasizing relational complementarity rather than fixed binaries. Health, in this context, is sustained through reciprocal responsibility rather than individual autonomy.

One of the most striking lessons of this chapter is the recognition of the land's own grief and continued presence. Healing does not require pristine or untouched landscapes. It requires sincerity, attentiveness, and willingness to remain in relationship even with damaged places. Scarred land continues to hold space for ceremony, instruction, and strength. Healing does not erase trauma—whether embodied in people or territory—but walks with it. Scars become maps of survival rather than evidence of failure, teaching resilience, accountability, and care (Whyte, 2020).

Finally, the chapter affirms that land-based healing has systemic and institutional implications. Practices such as compassionate listening within correctional spaces, resistance to capitalist pressures shaping youth wellbeing, and attention to environmental signs demonstrate that healing extends beyond individuals into the structures that govern daily life. In these contexts, listening becomes ceremony, compassion becomes medicine, and offering becomes a form of resistance to systems built on extraction and dispossession (Wilson & Inkster, 2018; McGregor, 2021). Land-based healing does not offer a step-by-step program or a universal model for replication. Instead, it offers a lifelong invitation to remember how to live in relationship—with land, with water, with one another, and with future generations. As demonstrated throughout this chapter, healing is not solely about fixing what is broken; it is about remembering what sacred and reorienting governance, sustainability, and care around that remembrance is.

REFERENCES

Craft, A., McGregor, D., & McKay, C. (2019). Indigenous legal traditions and water governance. *McGill Law Journal. Revue de Droit de McGill*, *64*(4), 755–790. DOI: 10.7202/1060866ar

Daigle, M. (2018). Resurging through Kishiichiwan: The spatial politics of Indigenous water relations. *Decolonization*, *7*(1), 159–172.

McGregor, D. (2021). Indigenous water governance and sustainability. In Renzetti, S., & Dupont, D. P. (Eds.), *Water policy and governance in Canada* (pp. 193–214). Springer., DOI: 10.1007/978-3-030-77502-4_9

McGregor, D., Whitaker, S., & Sritharan, M. (2020). Indigenous environmental justice and sustainability. *Current Opinion in Environmental Sustainability*, *43*, 35–40. DOI: 10.1016/j.cosust.2020.01.007

Simms, R., Harris, L. M., Joe, N., & Bakker, K. (2019). Navigating the tensions in collaborative watershed governance: Indigenous rights, participation, and authority. *Water Alternatives*, *12*(1), 1–22.

Smith, L. T. (2020). The native and the neoliberal down under: Neoliberalism and "endangered authenticities". In *Indigenous experience today* (pp. 333-352). Routledge.

Whyte, K. P. (2020). Indigenous climate change studies: Indigenizing futures, decolonizing the Anthropocene. *English Language Notes*, *57*(1), 153–162. DOI: 10.1215/00138282-7570055

Wilson, N. J., & Inkster, J. (2018). Respecting water: Indigenous water governance, ontologies, and the politics of kinship. *Environment and Planning. E, Nature and Space*, *1*(1–2), 223–242. DOI: 10.1177/2514848618789378

Wilson, S. (2008). Research is ceremony: Indigenous research methods. Fernwood Publishing. *(Still foundational; appropriate to cite alongside recent literature)*

Yates, J. S., Harris, L. M., & Wilson, N. J. (2017). Multiple ontologies of water: Politics, conflict and implications for governance. *Environment and Planning. D, Society & Space*, *35*(5), 797–815. DOI: 10.1177/0263775817700395

Chapter 10
Indigenous Futures:
Ceremony, Climate Resilience, and Water Sustainability Beyond Crisis

ABSTRACT

This chapter examines Indigenous water ceremonies as vital expressions of relational governance, showing how ceremony actively shapes sustainability practices, collective healing, and community resilience. Drawing on teachings and narratives shared by Indigenous knowledge holders, it situates water ceremony within ongoing contexts of colonization, environmental degradation, climate change, and cultural disruption. Rather than portraying ceremony solely as a response to crisis, the chapter emphasizes its continuous and anticipatory nature—grounded in everyday acts of respect, remembrance, and responsibility that sustain waterscapes across generations. Through this lens, water ceremony emerges not only as spiritual practice, but as governance in action.

INTRODUCTION

This chapter examines Indigenous water ceremonies as central expressions of relational governance, demonstrating how ceremony actively organizes sustainability practice, healing, and community resilience. Drawing on narratives, teachings, and ceremonial practices shared by Indigenous knowledge holders, the chapter situates water ceremony within broader contexts of colonization, environmental degradation, climate change, and cultural disruption. Rather than framing ceremony as a response to crisis alone, the chapter shows that Indigenous water governance is continuous

DOI: 10.4018/979-8-3373-7559-5.ch010

and anticipatory—rooted in daily acts of respect, remembrance, and responsibility that sustain waterscapes over time (Simpson, 2017; Todd, 2016).

Indigenous water knowledge constitutes a living system of relational governance that organizes sustainability practice, community health, and collective responsibility through ongoing relationships with water. Across Indigenous nations, water is understood not as a passive resource or symbolic cultural element, but as a living relative whose wellbeing is inseparable from that of land, people, and future generations. This relational ontology—often articulated through teachings such as *"water is life"*—positions water as a legal and ethical authority that guides behavior, shapes decision-making, and sustains ecological balance (Anderson et al., 2019; McGregor, 2014). Within this worldview, caring for water is not a discretionary environmental activity but a moral, spiritual, and governance obligation enacted through ceremony, daily practice, and intergenerational accountability. Recent scholarship in Indigenous water governance emphasizes that Indigenous legal orders and stewardship systems predate—and continue to operate alongside—colonial regulatory frameworks (Borrows, 2010; McGregor, 2021). These systems are grounded in reciprocal relationships rather than extractive management, privileging responsibility, humility, and long-term ecological continuity over control or commodification. Indigenous water governance is enacted through lived practices such as ceremonial offerings, water walks, seasonal observation, song, prayer, and collective monitoring—activities that function as operational mechanisms of governance rather than symbolic expressions alone (Chapola et al., 2025). Through these practices, communities regulate conduct, transmit ecological law, and maintain accountability to water as a sentient being.

This chapter examines Indigenous water ceremonies as central expressions of relational governance, demonstrating how ceremony actively organizes sustainability practice, healing, and community resilience. Drawing on narratives, teachings, and ceremonial practices shared by Indigenous knowledge holders, the chapter situates water ceremony within broader contexts of colonization, environmental degradation, climate change, and cultural disruption. Rather than framing ceremony as a response to crisis alone, the chapter shows that Indigenous water governance is continuous and anticipatory—rooted in daily acts of respect, remembrance, and responsibility that sustain waterscapes over time (Simpson, 2017; Todd, 2016).

The persistence of Indigenous water governance must be understood alongside the profound impacts of colonial water regimes. In Canada and other settler-colonial contexts, Indigenous communities continue to experience unsafe drinking water, industrial contamination, and exclusion from statutory decision-making processes despite holding inherent rights and responsibilities to water (McGregor et al., 2020; Norman et al., 2023). Long-term boil-water advisories, chemically treated water, and degraded fisheries represent not only infrastructural failures but relational ruptures that undermine trust, health, and spiritual wellbeing. Indigenous responses to these conditions—including water ceremonies, community advocacy, and land-based monitoring—represent acts of governance that challenge colonial systems while reaffirming Indigenous legal and ethical orders.

Importantly, Indigenous water ceremonies do not separate spiritual practice from environmental management. Ceremony operates as law-in-action, embedding sustainability within everyday life rather than confining it to institutional or technical domains. Practices such as offering tobacco at water's edge, singing to rivers, observing seasonal changes, and interpreting environmental signs function as regulatory processes that define ethical limits and guide adaptive responses to ecological change (Whyte, 2018). These practices establish obligations not only to current community members but to ancestors, water beings, and generations yet to come.

This chapter provides a clear translation layer for academic, policy, and practitioner audiences by demonstrating how Indigenous water knowledge functions in practice as environmental governance and sustainability action. Rather than reducing Indigenous knowledge to values or cultural insights that supplement Western science, the analysis foregrounds Indigenous water governance as a complete and coherent system with its own protocols, accountability structures, and decision-making processes. This approach responds directly to calls in sustainability scholarship for frameworks that integrate social justice, ecological integrity, and cultural continuity without reproducing colonial hierarchies of knowledge (Kimmerer, 2013; Todd, 2016).

The chapter also addresses the interconnected physical, emotional, spiritual, and ecological dimensions of water-related wellbeing. Indigenous teachings presented here illustrate how water ceremonies support healing from trauma, grief, illness, and climate-driven disruption by restoring relational balance. Drought, pollution, flooding, and seasonal instability are interpreted not solely as technical challenges but as communications from water itself—requiring attentive listening, humility, and relational adaptation rather than domination or control (Anderson et al., 2019).

Structurally, the chapter moves from community-based and ceremonial practices toward collective, institutional, and policy-relevant dimensions of water governance. This progression reveals how Indigenous water knowledge translates into adaptable sustainability pathways that can inform watershed planning, environmental assessment, and governance reform without collapsing into program-specific prescriptions. By centering ceremony, relational ethics, and intergenerational responsibility, Indigenous water governance offers guidance for rethinking sustainability as an ongoing practice of care rather than a crisis-driven intervention. Thus, this chapter positions Indigenous water ceremonies not as remnants of the past but as future-oriented governance systems capable of addressing contemporary environmental challenges. By grounding sustainability in sacred law, reciprocity, and accountability, Indigenous water governance resists commodification and technocratic abstraction.

It reframes sustainability as a lived moral practice—one that prioritizes ecological integrity, community wellbeing, and responsibility to future generations. In doing so, the chapter contributes to a growing body of work that recognizes Indigenous water knowledge as a leading framework for rethinking water sustainability in theory and in practice.

INDIGENOUS LAND-BASED SURVIVAL AS SACRED AND ESSENTIAL RELATIONSHIP WITH LAND AND WATER

Findings from this chapter demonstrate that Indigenous survival is not merely dependent on land and water as material resources, but is fundamentally grounded in sacred, protective, and reciprocal relationships with water-bearing lands. For Indigenous Peoples, survival is enacted through ongoing obligations to protect land and water systems—particularly muskegs, rivers, streams, and wetlands—that sustain ecological balance, cultural continuity, and collective wellbeing. These relationships are increasingly threatened by industrial extraction, environmental contamination, and colonial water governance regimes. The findings in this theme reveal how Indigenous communities respond to these threats through resistance, ceremony, and intergenerational responsibility, while simultaneously carrying grief for lands and waters already harmed.

Protection of Land from Industrial and Environmental Threats

A central finding is that protecting water-bearing lands from industrial extraction constitutes a core expression of Indigenous governance and survival. Participants consistently described resistance to strip mining, resource extraction, and land disturbance as necessary acts of ecological defense rooted in ancestral law rather than environmental activism alone. Muskegs, wetlands, and river systems are understood as living infrastructures that regulate water flow, filter contaminants, and sustain surrounding ecosystems. Their destruction threatens not only environmental health but spiritual balance and community survival. Women's leadership emerged as particularly significant in these protection efforts. Indigenous women were described as carrying responsibilities to water and land that position them at the forefront of resistance to industrial harm. One participant stated: "We work with a group of women who are noticing, trying to stop, strip mining of five areas, if those muskegs get strip-mined, it's going to create runoff into the rivers, streams, and lakes." This statement illustrates how Indigenous environmental governance is grounded in relational foresight rather than reactive mitigation. The protection of muskegs is not framed as opposition to development in abstract terms, but as a concrete obligation to prevent cascading ecological harm across watersheds. These findings demonstrate that Indigenous governance prioritizes prevention, relational accountability, and long-term ecological integrity over short-term economic benefit.

Ancestral teachings further reinforce this governance logic. Elders' warnings against disturbing the ground were repeatedly cited as guiding principles for land stewardship. One participant recalled: "The old people said that we shouldn't be taking things from the ground. Because if we take things from the ground, it will destroy everything around us."

This teaching reflects a deeply embedded understanding of ecological reciprocity and cumulative impact. Extraction is not viewed as an isolated act, but as a disruption that reverberates through land, water, animals, and human health. These findings show that Indigenous resistance to industrial extraction is not oppositional by nature; it is protective, preventative, and rooted in ceremonial law that defines acceptable relationships with land and water.

Collectively, these findings position Indigenous land protection as a form of water-centered governance that asserts sovereignty, enacts survival, and maintains ecological balance through relational responsibility rather than regulatory enforcement alone.

Challenges in Water Quality and Conservation

A second key finding concerns the deterioration of water quality resulting from colonial interventions and externally imposed treatment systems. Participants described how state-controlled water management strategies—particularly chemical treatment and boil-water advisories—have eroded trust in water systems while introducing new health risks. Rather than restoring safety or confidence, these interventions often compound harm by disconnecting communities from their water sources and undermining Indigenous water knowledge. One participant explained: "Our health department puts chlorine in our water to meet testing thresholds, then issues boil-water advisories. Many in our community refuse chlorinated water—diabetes, cancer, lung issues are rampant." This finding highlights how colonial water governance prioritizes compliance with technical standards over relational health outcomes. Chemical treatment may satisfy regulatory benchmarks, yet it fails to address community wellbeing or restore trust in water as a living relative. The refusal of chlorinated water is not ignorance or resistance to science; it reflects lived experience of harm and a relational understanding of water as something that must remain life-giving.

Intergenerational observations further illustrate the visible degradation of water quality. One participant recalled: "When I was a child, if you cooked fish in a pot, you just cleaned the pot. Now you have to scrub it because the water is so dirty."

This statement conveys mourning for the loss of water purity and signals the cumulative effects of pollution over time. The findings show that water degradation is not abstract or invisible; it is experienced daily through altered taste, residue, illness, and loss of trust. Indigenous communities are therefore navigating a damaged system in which water governance has been severed from relational accountability. Despite these challenges, participants emphasized continued efforts toward community-based stewardship and ceremony as pathways for restoring water relationships. These findings demonstrate that water ceremony functions not only as spiritual practice but as a response to governance failure—reasserting Indigenous authority, care, and responsibility in the face of imposed systems that have compromised water integrity.

INDIGENOUS LAND-BASED CHALLENGES AS ENVIRONMENTAL DECLINE AND THE SACREDNESS OF WATER

Findings from Theme 2 reveal that environmental decline is interpreted by Indigenous communities as both an ecological and spiritual crisis. Changes in water systems—such as drying rivers, disappearing springs, and declining fish populations—are understood as signs of imbalance that extend beyond physical degradation. These changes signal ruptures in sacred relationships between humans, water, and the Earth, requiring ceremonial, relational, and collective responses.

Water as Life and the Blood of Mother Earth

A recurring finding across participant narratives is the conceptualization of water as the blood of Mother Earth. This metaphor situates water within a living body, emphasizing its role in sustaining life and regulating vitality across ecosystems. Participants described the loss of water flow as analogous to a body losing circulation—an indicator of deep systemic illness. One participant expressed this connection poignantly: "We used to drink straight from springs—our mother's blood. Now those springs have dried up, fingers wither when blood stops, and so does the land." This metaphor underscores how water loss is experienced as both environmental and spiritual trauma. The drying of springs is not merely hydrological change; it represents a breakdown in life-sustaining relationships. Findings show that Indigenous cosmologies frame water harm as a wound to the Earth itself—an injury that demands ceremonial response, not technical repair alone. This understanding reinforces the sacredness of water and the moral responsibility humans carry to protect its flow, purity, and vitality. Damaging water systems is understood as an act that reverberates through all forms of life, requiring accountability at spiritual, communal, and governance levels.

Urgent Signs of Ecological Collapse

A final key finding concerns the interpretation of environmental signs as urgent calls to action. Participants described how drying streams, disappearing fish, altered seasons, and ecosystem instability are read through generations of land-based observation. These signs are not treated as anomalies but as confirmations of long-anticipated ecological decline resulting from unsustainable extraction and disrupted relationships. One participant noted: "Streams are drying up; the fish we used to see every spring haven't come for years." This observation reflects ecological grief as well as foresight. Indigenous communities have long warned that disrupting land and water systems would lead to collapse. The absence of fish and water flow confirms these teachings and signals the need for immediate relational intervention. These findings demonstrate that Indigenous environmental monitoring is inseparable from ceremony and observation. Knowledge of ecological change is embedded in seasonal rhythms, species behavior, and water movement rather than abstract data alone. This form of governance integrates spiritual attentiveness with empirical awareness, reinforcing that sustainability requires listening to land and water as communicative beings.

Together, these themes demonstrate that Indigenous survival, governance, and sustainability are enacted through sacred relationships with land and water. Protection of water-bearing lands, resistance to extraction, grief over water degradation, and interpretation of ecological signs all function as governance practices grounded in relational accountability. These findings position Indigenous water knowledge not as supplementary to sustainability discourse, but as a foundational framework for protecting ecosystems, sustaining life, and responding ethically to environmental decline. Indigenous land- and water-based governance is therefore best understood as survival in practice—rooted in responsibility, ceremony, and intergenerational care rather than control or exploitation.

INDIGENOUS VIEWS OF URBAN DISCONNECTION AND ENVIRONMENTAL DEATH

Findings from this theme reveal a consistent Indigenous understanding of urban environments as sites of profound spiritual and ecological disconnection. Cities are not described merely as densely populated spaces, but as landscapes that have been severed from the sacred rhythms of water, land, and life. Concrete infrastructure, industrial noise, and the concealment or degradation of water systems contribute to an experience of lifelessness that undermines relational wellbeing. Within Indigenous epistemologies, water must be visible, accessible, and relational to sustain spiritual balance and ethical responsibility. When water is buried, channelized, or rendered inaccessible, relationships with life itself are disrupted.

Participants articulated urban disconnection not as an abstract critique of modernity, but as a lived emotional and spiritual experience marked by grief, suffocation, and alienation. Urban spaces were repeatedly contrasted with water-rich land-based environments where relationships with rivers, springs, lakes, and wetlands support ceremonial practice and daily renewal. The findings indicate that cities are experienced as environments where Indigenous governance relationships with water are structurally obstructed.

Cities as Places of Lifelessness

A central finding is that cities are perceived as spiritually barren landscapes where life-giving relationships with water are largely absent. One participant described this condition starkly: "Now cities are of dead rock, metal and cement; nothing lives there." This statement reflects not only environmental degradation, but a deeper sense of spiritual deprivation. In Indigenous worldviews, life is sustained through reciprocal relationships with water, land, plants, and non-human beings. Urban environments—where water is hidden underground, polluted, or commodified—interrupt these relationships and prevent the enactment of ceremonial responsibility. The absence of visible water and living ecosystems is experienced as a form of environmental death.

The findings suggest that urbanization represents a broader rupture in Indigenous governance systems, as cities are built on principles of extraction, separation, and control rather than reciprocity. Water becomes infrastructure rather than relative, managed through pipes and regulations rather than ceremony and relationship. This disconnection has direct implications for Indigenous wellbeing, as the inability to engage with water relationally contributes to spiritual exhaustion and cultural dislocation.

Importantly, participants emphasized that the remedy for urban lifelessness is not abandonment of cities alone, but the restoration of water-centered ceremonial practices that can reanimate relationship even in damaged environments. Returning to water ceremonies—wherever water remains accessible—is understood as a means of resisting environmental death and restoring vitality in spaces shaped by colonial development.

COLLECTIVE RESPONSIBILITY AND ADVOCACY AS WATER GOVERNANCE

Findings from this theme demonstrate that Indigenous water protection is fundamentally collective, embedded in shared responsibility, and enacted through both public advocacy and everyday ceremonial practice. Water governance is not delegated to specialists or institutions alone; it is a communal obligation sustained through education, ceremony, and intergenerational participation. These findings reveal that Indigenous governance operates through distributed accountability, where every community member holds responsibility for water care.

Community Awareness and Collective Action

A key finding is that water protection is understood as a shared covenant rather than an individual or professional task. Participants described governance structures that involve Elders, youth, boards, councils, educators, and families working together to protect water systems. One participant articulated this collective ethic clearly: "It's not just four or five staff—our board, our Elders Council, it's everyone's responsibility to take care of water." This statement reflects a governance model grounded in relational accountability rather than hierarchical authority. Water advocacy is not separated from daily life or ceremonial responsibility; it is woven into how communities organize, teach, and act together. The findings show that Indigenous water governance is inherently participatory, requiring collective awareness and shared action across generations.

Community-based water advocacy includes monitoring environmental change, resisting harmful development, educating youth, and maintaining ceremonial relationships with water. These practices reinforce that sustainability is not achieved through regulation alone, but through collective commitment and ethical obligation. Water protection is simultaneously political, spiritual, and relational—bridging advocacy with ceremony rather than separating them.

Everyday Acts of Prayer and Respect

In addition to collective action, the findings highlight the significance of everyday, individual acts of respect as forms of water governance. Participants emphasized that ceremony does not require formal gatherings or public recognition; intention itself carries power. One participant explained: "Even alone at the water's edge you can offer a silent prayer—water hears and responds." This finding underscores that Indigenous water governance operates at multiple scales, from institutional advocacy to intimate personal practice. Silent prayer, offerings, and mindful presence at the water's edge are understood as acts of accountability that reaffirm water's agency and relational status. These everyday practices sustain ethical relationships and reinforce water's role as a conscious being rather than an inert resource. The findings suggest that such acts are not symbolic gestures but operational mechanisms of governance. They cultivate humility, attentiveness, and responsibility, reinforcing the moral foundations of sustainability. Through these practices, individuals actively participate in maintaining balance, even in the absence of formal authority or ceremony.

LAND-BASED CHILDHOOD MEMORIES AND THE FORMATION OF WATER RELATIONSHIPS

Findings from this theme shows that relationships with water are often formed in early childhood through experiences of wonder, emotional resonance, and sacred curiosity. These formative encounters shape lifelong orientations toward land, healing, and ceremony. Water relationships are not learned solely through instruction; they emerge through embodied experience, sensory engagement, and storytelling. Childhood memories of water were described as foundational moments in which spiritual awareness and relational understanding first take root. These early experiences serve as the basis for later responsibilities to protect water and uphold ceremonial relationships.

Wonder, Emotion, and Formative Experiences

A key finding is that childhood interactions with water often carry deep emotional and spiritual significance. Participants described early memories marked by curiosity, beauty, and a sense of mystery. One participant reflected: "As a child I wondered why water looks different day vs. night—everything is a wonder." This statement illustrates how water teaches through observation and presence. Changes in light, movement, and reflection are not trivial details; they invite attention, reverence, and relational engagement. The findings show that such moments cultivate sensitivity to water as a living presence rather than a static object. These early experiences are not purely personal; they function as sites of cultural transmission. Stories about water, shared by family members and Elders, reinforce the sacredness of these encounters and embed them within collective memory. Childhood wonder becomes the root from which ceremonial responsibility and governance grow. Through these formative experiences, individuals learn to see water as teacher, relative, and guide.

These themes demonstrate that Indigenous water governance is shaped by lived experience across environments, generations, and scales of action. Urban disconnection reveals the consequences of severed relationships with water, while collective advocacy and everyday ceremony illustrate how responsibility is reclaimed and enacted. Childhood wonder underscores that sustainability begins not with policy, but with relationship. These findings reinforce the book's central argument that rethinking water sustainability requires restoring relational governance—grounded in visibility, responsibility, and lived connection with water. Indigenous water knowledge offers not only critique of environmental decline, but pathways for renewal rooted in collective care, daily practice, and intergenerational memory.

LAND-BASED BALANCE, REVERENCE, AND SUSTAINABILITY

Findings from this theme demonstrate that Indigenous sustainability is grounded not in regulation alone, but in daily practices of reverence, gratitude, and ceremonial alignment with water and the natural world. Balance is not understood as a static ecological condition; it is actively maintained through ritual action, song, and seasonal teachings that guide human behavior in relation to water. Gratitude, in this context, functions as a governance mechanism—shaping ethical conduct, reinforcing humility, and sustaining reciprocal relationships between people and the more-than-human world. Ceremonial practices centered on thankfulness affirm that water is not an inert substance but a living presence requiring acknowledgment and care. These practices embed sustainability into everyday life, ensuring that ecological responsibility is not deferred to institutions but enacted continually through relational obligation.

Learning Gratitude through Ceremony and Song

A central finding is that gratitude is cultivated through embodied ceremonial practice rather than abstract moral instruction. Singing to water and the elements is not symbolic performance; it is an act that aligns human intention with ecological order. One participant explained:

"We sing songs to water and all four elements—fire, sun, water, earth—to live in balance and express our thanks every day." This statement reflects a worldview in which sustainability is achieved through daily acknowledgment of interdependence. Song functions as both teaching and practice, reinforcing humility and attentiveness to ecological relationships. By singing to water, individuals reaffirm their responsibility to live in ways that do not disrupt balance.

The findings suggest that gratitude ceremonies regulate behavior by cultivating awareness of dependence on water systems. Rather than relying on enforcement, these practices foster internalized accountability grounded in reverence. Sustainability, in this framework, is not a technical solution but a lived ethic continuously renewed through ceremony.

THE KEEPERS MOVEMENT—ORIGINS AND GROWTH

Findings from this theme illustrate how Indigenous water governance extends from ceremony into organized collective action. The Keepers of the Water movement emerged as both a spiritual and political response to escalating water crises, demonstrating how Indigenous governance systems mobilize ceremony, science, and advocacy without fragmenting these domains. The movement exemplifies how relational governance can scale outward from sacred declaration to institutional engagement. Rather than separating healing from activism, the Keepers movement integrates both, recognizing that protecting water requires restoring relationships among people as much as addressing environmental threats.

Formation in Response to Water Crisis

Participants described the formation of the Keepers of the Water movement as a response rooted in ceremony rather than reactionary politics. A foundational moment involved declaring water sacred, which then catalyzed broader organizing efforts. One participant recalled: "In 2006 we declared water sacred and formed Keepers of the Water. Gathering together—Elders, youth, scientists—is itself deeply healing."

This quote highlights a key finding: gathering itself is a form of healing governance. Bringing together diverse knowledge holders affirms that water protection requires intergenerational wisdom, scientific insight, and spiritual responsibility. Ceremony does not precede activism; it constitutes it.

The findings show that the movement's strength lies in its refusal to separate legal resistance from spiritual obligation. Water is defended not only through policy advocacy but through continued ceremonial relationship, ensuring that activism remains accountable to sacred law rather than political expediency.

INDIGENOUS LAND-BASED HEALING, CEREMONY, AND CONNECTION

This theme 8 findings emphasize that gathering in ceremony around water is a powerful mechanism for collective healing and relational restoration. Water ceremonies create spaces where grief, responsibility, and renewal can be shared, reinforcing community bonds and sustaining ethical commitment to water protection. These gatherings function as governance sites where relational accountability is renewed through presence rather than instruction.

Gathering as a Form of Healing

A key finding is that ceremony does not require speech or formal ritual to be effective. The act of standing together with shared intention carries profound healing power. One participant described this experience: "Just being able to stand in a circle, even in silence, and know we are offering to the water—that was enough." This statement reveals that healing emerges from collective presence and mutual recognition of responsibility. Silence, in this context, is not absence but attentiveness. Gathering allows participants to process grief related to environmental loss while reaffirming commitment to care for water. The findings indicate that such ceremonies provide emotional regulation and spiritual grounding, reinforcing that sustainability is inseparable from collective wellbeing. Healing water and healing community are understood as simultaneous processes.

Relationship Building with Diverse Groups

Another important finding is that Indigenous-led water ceremonies foster relationships across cultural and generational boundaries. Participants emphasized that ceremony is an educational space where youth and non-Indigenous allies learn how to listen to water respectfully. One participant noted: "It's not just us—the youth, the allies, they show up to learn how to listen to the water too." This highlights ceremony as a site of relational learning rather than instruction. Allies are not positioned as saviors or experts but as listeners invited into ethical relationship. The findings suggest that such inclusive ceremonies expand networks of responsibility while maintaining Indigenous leadership and protocol. Through these gatherings, water governance becomes a shared commitment that extends beyond Indigenous communities without diluting Indigenous authority or sacred law.

INDIGENOUS LAND-BASED SACRED BEINGS AND THE WATER REALM

Findings from this theme reveal that spiritual beings and mythic narratives play a critical role in guiding ethical relationships with water. Stories of water beings are not folklore detached from governance; they function as instructional frameworks that teach humility, attentiveness, and care. These narratives encode protocols for listening and responding to water as a conscious presence.

The Mythical Baby and Listening to Elders

Participants described water beings as reminders of water's agency and vulnerability. Stories about mythic figures, such as the baby in the water, serve as ethical teachings passed down through oral tradition. One participant explained: "The baby in the water isn't just a story—it's a reminder that we must always listen when water speaks." This finding underscores that listening is a core governance principle. Water communicates through signs, stories, and ecological change, and humans are obligated to respond with care rather than control. Listening to Elders ensures that interpretations of these teachings remain grounded in lived experience and relational accountability. The findings demonstrate that mythic narratives sustain water ethics across generations, reinforcing humility and restraint. These stories regulate behavior by reminding communities that water must be approached with respect, patience, and reverence.

Together, these findings demonstrate that Indigenous water sustainability is enacted through reverence, ceremony, movement-building, collective gathering, and spiritual listening. Balance is maintained not through domination of water systems, but through gratitude, shared responsibility, and attentiveness to sacred teachings. From daily songs of thanks to organized movements like the Keepers of the Water, Indigenous governance systems integrate healing, activism, and ethics into cohesive sustainability practice. These themes reinforce the book's central argument: rethinking water sustainability requires restoring relational governance grounded in humility, ceremony, and collective responsibility. Indigenous water knowledge offers not only critique of environmental degradation, but enduring pathways for renewal rooted in listening, gathering, and honoring water as a living relative.

LAND-BASED ORIGINS AND PERSONAL JOURNEY WITH WATER CEREMONY

Findings from this theme demonstrate that ceremonial leadership in Indigenous water governance does not emerge through individual ambition or institutional training, but through lineage, relational accountability, and sustained mentorship within families and communities. Becoming a ceremonial leader is a gradual process shaped by memory, responsibility, and intergenerational instruction. Water knowledge is not acquired; it is carried, and only when a person has been prepared—emotionally, spiritually, and relationally—by those who came before them. This theme highlights that water ceremony is governed through kinship structures that function as systems of authorization and ethical regulation. Family teachings establish boundaries, transmit protocol, and ensure continuity of responsibility across generations. In this way, ceremonial leadership becomes an expression of Indigenous governance rather than personal identity.

Family Teachings and Ceremonial Lineage

Participants consistently emphasized that water knowledge is passed through family relationships, not through formalized or external systems. Ceremonial readiness is cultivated through daily observation, mentorship, and responsibility modeled by relatives who already carry the work. One Indigenous storyteller explained: "My dad and my auntie Diana—they were the ones who prepared me to carry this water work." This statement illustrates how family members act as primary governance bodies in water ceremony. Preparation is not abstract instruction but lived apprenticeship—watching how offerings are made, how water is spoken to, how humility is practiced, and how responsibility is carried over time. The findings indicate that family lineage serves as a safeguard against misuse of ceremony. Grounding water teachings in kinship, communities ensure that knowledge remains accountable to place, memory, and relationship rather than detached from context. Lineage-based transmission preserves both protocol and ethical restraint, reinforcing water ceremony as a collective responsibility rather than an individual platform.

Personal Transformation and Becoming a Kôkum

Ceremonial leadership also requires personal transformation marked by life events that signal readiness and responsibility. Becoming a kôkum—a ceremonial grandmother—is not simply a title but a recognition of spiritual maturity and relational accountability. One participant shared: "Two years ago I became a kôkum—and I led my first water ceremony the day my son was born." This quote reveals how ceremonial responsibility is interwoven with life transitions that deepen relational awareness. Birth, loss, and caregiving are not separate from governance; they shape one's capacity to carry water work with humility and care.

The findings suggest that water ceremony requires alignment between personal readiness and communal trust. Leadership is acknowledged when an individual has demonstrated patience, emotional steadiness, and commitment to collective wellbeing. In this sense, ceremonial roles are not claimed but bestowed, ensuring continuity of ethical governance within water stewardship.

INDIGENOUS LAND-BASED CONNECTION TO WATER AND HONORING GRANDMOTHERS

This theme findings highlight that Indigenous water governance is sustained through disciplined daily practice rather than occasional ceremony. Relationships with water are renewed through consistent acts of gratitude, offering, and respectful conduct that honor grandmother teachings and matriarchal responsibility. These everyday practices function as micro-governance actions that reinforce accountability, humility, and care. Honoring grandmothers—both human and spiritual—is central to maintaining ethical relationships with water. Through repetition and presence, individuals learn that sustainability is not event-based but enacted through routine relational commitment.

Daily Offerings and Personal Practice

Participants described daily offerings to water as foundational to their ceremonial responsibility. These acts are not symbolic gestures but reaffirmations of relational obligation. One individual explained: "I make an offering to the water every morning—before I do anything else." This practice demonstrates that water governance begins with prioritization. Offering first establishes a rhythm that centers responsibility to water above personal convenience or productivity. The findings indicate that daily practice cultivates attentiveness and prevents relational neglect.

Such practices ensure that water is consistently acknowledged as a living relative rather than an assumed resource. Over time, these routines shape ethical behavior, reinforcing long-term stewardship through habit rather than enforcement.

Respect and Everyday Ceremony

Another key finding is that ceremony is embedded in everyday interactions with water, not limited to formal gatherings. Participants emphasized that how water is handled reflects one's relationship to it. As one speaker noted: "Ceremony isn't only at gatherings. It's in how you pour, how you carry, how you thank." This quote illustrates that ethical water governance is enacted through small, continuous actions. Respectful conduct becomes a daily protocol that reinforces sacred law. By embedding ceremony in routine behavior, Indigenous governance systems maintain consistency and accountability across all scales of interaction.

INDIGENOUS TRADITIONAL INNER HEALING AND WATER

This theme explains that water healing is reciprocal and contingent upon emotional readiness. Water does not act upon individuals unilaterally; healing requires openness, vulnerability, and willingness to release pain. Inner preparation is therefore a critical governance mechanism that regulates access to healing and ensures respectful engagement.

Self-Cleansing and Emotional Readiness

Participants emphasized that water responds to intention and emotional presence. Healing is not automatic but activated through personal readiness. One participant explained: "You have to be ready. The water doesn't just take your pain—you have to let it go." This finding underscores that water healing involves accountability to one's own emotional state. Ceremony creates the conditions for healing, but individuals must actively participate by surrendering what no longer serves them.

The findings suggest that emotional readiness functions as a boundary that prevents superficial or extractive engagement with ceremony. Healing occurs when individuals approach water with honesty and humility, reinforcing water's role as a relational agent rather than a passive tool.

INDIGENOUS LAND-BASED SACRED GRANDMOTHERS AND CEREMONY HELPERS

This theme findings affirm that water ceremonies are supported by spiritual governance structures embodied by grandmother spirits and their helpers. These beings are not abstract symbols but active participants in ceremonial life, guiding prayer, protecting protocols, and maintaining balance. Honoring grandmother spirits reinforces matriarchal leadership and relational ethics within water governance systems.

Grandmother Spirits and Their Roles

Participants described grandmother spirits as organizing figures within ceremonial cosmology. One explained: "They say there are four grandmother beings—and each has helpers, just like we do." This statement highlights that spiritual governance mirrors human structures of responsibility and care. Grandmother spirits embody wisdom, patience, and protection, ensuring that ceremonies remain aligned with sacred law. The findings indicate that recognizing these spiritual roles sustains ethical continuity across generations. By honoring grandmother spirits, communities reaffirm the spiritual architecture that governs water ceremonies, ensuring that healing and stewardship remain relational, respectful, and accountable.

These themes reinforce the book's central argument that rethinking water sustainability requires attention to relational processes that cannot be legislated but must be lived. Indigenous water ceremonies function as systems of governance that regulate behavior, transmit ethical law, and sustain ecological and communal wellbeing across generations.

LEARNING REFLECTION

This chapter has demonstrated that Indigenous water ceremonies function as comprehensive systems of relational governance that sustain cultural identity, community resilience, and environmental stewardship. Rather than existing as symbolic or isolated cultural rituals, water ceremonies operate as living frameworks of law, healing, ethics, and decision-making that guide how Indigenous Peoples relate to water across generations (Borrows, 2010; McGregor, 2014). From Elder teachings to contemporary water protector movements, water emerges throughout this chapter as a spiritual, political, and ecological foundation for Indigenous life.

Across the thirteen themes and nineteen subthemes examined, one central finding is consistent: water is not a passive resource to be managed but a living relative that actively participates in healing, governance, and collective responsibility. This worldview directly challenges Euro-Western water governance frameworks that frame water primarily as property, infrastructure, or economic input (Simms et al., 2016). Indigenous water teachings instead emphasize reciprocity, relational accountability, and ethical restraint, affirming that water listens, responds, remembers, and requires care (Craft, 2017).

Many of the teachings shared in this chapter begin with grief. Participants described the emotional and spiritual pain of witnessing drying rivers, disappearing muskegs, polluted lakes, and contaminated fishing grounds. These losses are not merely environmental; they are profound relational ruptures that affect physical health, mental wellbeing, and cultural continuity (Whyte, 2017). Colonially imposed water treatment systems, industrial extraction, and urbanization have disrupted longstanding relationships between people and water, often introducing new forms of illness, distrust, and disconnection (McGregor et al., 2020). In Indigenous frameworks, harm to water is inseparable from harm to community.

At the same time, this chapter reveals that Indigenous water ceremonies are enduring, adaptive, and future-oriented. They are not static traditions preserved unchanged from the past; they are living practices that respond dynamically to contemporary environmental and political conditions (Coulthard, 2014). Women, grandmothers, ceremonial helpers, and youth continue to carry water teachings through daily offerings, songs, water walks, and collective gatherings. Healing practices—whether bathing, fasting, crying, praying, or listening—operate at both inner and communal levels, allowing emotional release while reaffirming relational responsibilities to water and land.

A key insight emerging from this chapter is that water ceremonies are inherently communal and political. Movements such as Keepers of the Water, alongside community-led rainwater harvesting, water monitoring, and ceremonial advocacy, demonstrate how Indigenous governance integrates scientific tools with spiritual and cultural knowledge (Arsenault et al., 2019). The use of pH meters alongside ancestral signs, or legal advocacy alongside prayer, does not represent contradiction but coherence within Indigenous epistemologies. These practices resist colonial binaries that separate science from spirituality and affirm that Indigenous knowledge systems are complete, adaptive, and capable of addressing contemporary sustainability challenges (Kimmerer, 2013; Simpson, 2017).

Another critical learning from this chapter is that ceremony is embedded in everyday life. Formal gatherings are not the only sites of ceremonial action. Small, consistent acts—offering water, speaking gratitude, singing quietly at a stream, or carrying water respectfully—are equally significant. This expands the definition of ceremony beyond event-based rituals and emphasizes that relational accountability is continuous rather than occasional. Everyday ceremonial conduct becomes the foundation for long-term ecological care and sustainability, reinforcing that restoration begins with daily practice.

The relational worldview that underpins Indigenous water ceremonies offers a profound alternative to extractive sustainability models. Where dominant frameworks emphasize efficiency, control, and consumption, Indigenous approaches center kinship, responsibility, and renewal (Whyte, 2020). Water is understood as the blood of the Earth, sustaining all life and connecting generations. To act in relation to water is therefore to act in relation to all beings—human and more-than-human alike. Sustainability, in this context, is not an abstract goal but a lived ethical commitment.

Throughout the chapter, the role of grandmothers and feminine spiritual leadership emerges as especially significant. Grandmother spirits, women water carriers, and Elders guide ceremonial protocols through listening, patience, and care rather than authority imposed through dominance or hierarchy (Craft, 2017). These roles are not symbolic but grounded in millennia of practice that encode governance principles through story, song, and ceremony. The power of these leaders lies not in speaking for water but in teaching others how to listen to it—an essential skill for ethical environmental stewardship.

Importantly, this chapter affirms that Indigenous water ceremonies are not backward-looking or nostalgic. They are future-making practices that model how communities can live well amid climate uncertainty, environmental degradation, and political marginalization (Whyte, 2017). Ceremony becomes a form of resistance—not through confrontation alone, but through continuity, care, and refusal to abandon relational responsibilities. In this sense, resistance itself becomes ceremonial, sustaining what colonial systems have attempted to erase.

Finally, this chapter affirms that healing is not only possible but already occurring. Healing unfolds each time a grandmother offers water to a river, each time a child sings to a lake, and each time a community chooses relational stewardship over state neglect. Healing takes place in silence and in song, in grief and in renewal. Indigenous water ceremonies demonstrate that sustainability is not something to be engineered in the future; it is practiced in the present through relationship, memory, and spirit.

To follow water, as this chapter has shown, is to follow ceremony, responsibility, and relational governance. Indigenous water knowledge does not merely contribute to sustainability discourse—it fundamentally reshapes it. By centering water as a living relative and governance partner, Indigenous water ceremonies offer pathways toward ecological care that are ethical, resilient, and deeply human.

REFERENCES

Anderson, K., Clow, B., & Haworth-Brockman, M. (2019). Carriers of water: Aboriginal women's experiences, relationships, and reflections. *Journal of Indigenous Wellbeing: Te Mauri-Pimatisiwin*, *4*(1), 1–19.

Arsenault, R., Bourassa, C., Diver, S., McGregor, D., & Witham, A. (2019). Including Indigenous knowledge systems in environmental assessments: Restructuring the process. *Global Environmental Politics*, *19*(3), 120–140. DOI: 10.1162/glep_a_00519

Borrows, J. (2010). *Canada's Indigenous Constitution*. University of Toronto Press.

Chapola, J., Datta, R., Starlight, T., Hurlbert, M., & Poggendorf, S. (2025). Indigenous women-led climate crisis solutions from decolonial feminist perspectives in Western Canada. *Environmental Science & Policy*, *174*, 104272. DOI: 10.1016/j.envsci.2025.104272

Coulthard, G. S. (2014). *Red skin, white masks: Rejecting the colonial politics of recognition*. University of Minnesota Press. DOI: 10.5749/minnesota/9780816679645.001.0001

Craft, A. (2017). *Breathing life into the Stone Fort Treaty: An Anishinabe understanding of treaty one*. UBC Press.

Kimmerer, R. W. (2013). *Braiding sweetgrass: Indigenous wisdom, scientific knowledge, and the teachings of plants*. Milkweed Editions.

McGregor, D. (2014). Traditional knowledge and water governance: The ethic of responsibility. *Alternative*, *10*(5), 493–507. DOI: 10.1177/117718011401000505

McGregor, D. (2021). Indigenous water governance in Canada. In Renzetti, S., & Dupont, D. P. (Eds.), *Water policy and governance in Canada* (pp. 87–106). Springer.

McGregor, D., Whitaker, S., & Sritharan, M. (2020). Indigenous environmental justice and sustainability. *Current Opinion in Environmental Sustainability*, *43*, 35–40. DOI: 10.1016/j.cosust.2020.01.007

Norman, E. S., Bakker, K., & Cook, C. (2023). Water governance and Indigenous rights in settler states. *Water International*, *48*(1), 1–20. DOI: 10.1080/02508060.2022.2147365

Simms, R., Harris, L. M., Joe, N., & Bakker, K. (2016). Navigating the tensions in collaborative watershed governance: Indigenous peoples, settler colonialism, and water justice. *Human Geographies*, *9*(2), 38–55.

Simpson, L. B. (2017). *As we have always done: Indigenous freedom through radical resistance*. University of Minnesota Press. DOI: 10.5749/j.ctt1pwt77c

Todd, Z. (2016). An Indigenous feminist's take on the ontological turn. *Journal of Historical Sociology*, *29*(1), 4–22. DOI: 10.1111/johs.12124

Whyte, K. (2017). Indigenous climate change studies: Indigenizing futures, decolonizing the Anthropocene. *English Language Notes*, *55*(1–2), 153–162. DOI: 10.1215/00138282-55.1-2.153

Whyte, K. (2020). Too late for indigenous climate justice: Ecological and relational tipping points. *Wiley Interdisciplinary Reviews: Climate Change*, *11*(1), e603. DOI: 10.1002/wcc.603

Whyte, K. P. (2018). Indigenous science (fiction) for the Anthropocene. *Environment and Planning. E, Nature and Space*, *1*(1–2), 224–242. DOI: 10.1177/2514848618777621

Land–Based Camps as a Transformative Method

IMAGES

Figure 1. Research team learning from elders

This image captures a relational learning space where Elders guide Indigenous and non-Indigenous researchers through collective dialogue. It reflects intergenerational knowledge exchange and the ethical foundations of Indigenous community-led water sustainability at Kâniyasihk Cultural Camps.

Figure 2. Individual indigenous story sharing

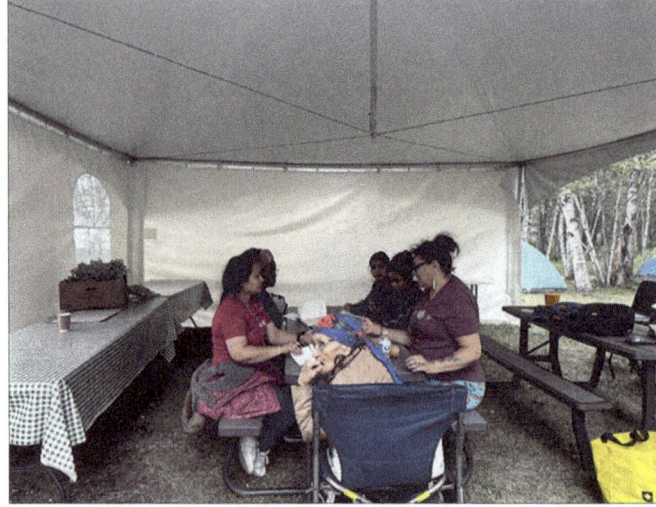

This photo highlights an individual knowledge holder sharing land-based teachings on water sustainability. It reflects how Indigenous storytelling functions as science, governance, and methodology, grounded in lived experiences, memory, and relational accountability with land and water.

Figure 3. Collective reflection after land-based camp

This image represents a shared moment of reflection following four days of land-based learning. Participants collectively express experiences, emotions, and teachings, demonstrating how knowledge emerges through relational engagement, ceremony, and community dialogue.

Figure 4. Collective land-based story sharing for water governance

This photo illustrates how collective storytelling on the land shapes Indigenous water governance. Stories shared within this circle function as living law, guiding responsibilities, ethics, and sustainable practices rooted in relational accountability.

Figure 5. Collective story sharing on community-led water sustainability

This image captures a collaborative storytelling circle where community members and researchers engage in dialogue about water. It reflects how shared narratives strengthen community-led governance, reinforce cultural teachings, and sustain relational responsibilities to water.

Figure 6. Indigenous art-based climate learning

This photo presents art-based activities as a form of Indigenous knowledge expression. Through creative practices, participants communicate understandings of the climate crisis, linking emotional, cultural, and ecological dimensions of environmental change.

Figure 7. Elder teaching land-based water stories (Saskatchewan Delta)

This image shows an Elder from the Cumberland Cree First Nation sharing teachings about water in the Saskatchewan Delta. It reflects how land-based stories transmit ecological knowledge, governance principles, and intergenerational responsibilities toward water.

Figure 8. Children's art and land-based learning

This photo highlights children engaging in art as part of land-based education. It demonstrates how youth learn relational connections to land and water through creative expression, fostering early understanding of responsibility and sustainability.

Figure 9. Elder-led water stories on community sustainability

This image captures an Elder leading storytelling on water sustainability. It emphasizes how Elders' teachings guide community-led approaches to water governance, embedding cultural, spiritual, and ecological responsibilities within everyday practices.

Figure 10. Elders and land-based activists on the meaning of water

This photo brings together Elders and land-based activists discussing Indigenous meanings of water. It reflects the collective articulation of water as a living relative, emphasizing responsibility, protection, and resistance to extractive systems.

Figure 11. Indigenous women teaching water and medicinal plants

This image highlights Indigenous women sharing knowledge about water and medicinal plants. It reflects matriarchal leadership in sustaining ecological knowledge, health practices, and cultural continuity within Indigenous water governance systems.

Figure 12. Semi-group land-based story sharing

This photo shows a smaller group engaged in land-based storytelling. It represents intimate knowledge exchange, where participants listen, reflect, and build relationships, reinforcing storytelling as both methodology and governance practice.

Glossary

Indigenous Knowledge Terminologies

Ceremonial Governance: Governance rooted in ceremony, spirituality, offerings, songs, and protocols. The book positions ceremony as scientific, political, legal, and ecological knowledge simultaneously.

Indigenous Legal Orders: Community-based laws carried through stories, ceremony, land practices, and kinship obligations—not colonial written policy.

Indigenous Traditional Land-Based Knowledge (ITLBK): A comprehensive worldview and methodology that understands land, water, air, and all beings as living relatives connected through reciprocal responsibilities. The book defines ITLBK as more than environmental observation—it is a relational worldview shaping governance and sustainability.

Land-Based Learning / Land-Based Education: Teachings received directly from participating in land activities such as ceremony, harvesting, storytelling, water walking, energy sovereignty camps, and seasonal knowledge. The book treats land as a teacher and governance source.

More-than-Human Relations: A term describing kinship with animals, plants, water, air, ancestors, and spiritual beings. Environmental decisions must honor these relations as living entities, not "natural resources."

Relational Accountability: A foundational ethical commitment describing how humans must behave toward land, water, ancestors, and more-than-human kin. Sustainability arises through accountable relationships, not technocratic management.

Story-Sharing / Storytelling as Method: Stories are teachings, governance, evidence, and law. Story-sharing is a research methodology, a land-based technology, and a healing practice.

Water Governance Terminologies

Ceremonial Water Walks: Community-led walks to honor, protect, and re-establish responsibility for water. These walks are governance, education, and healing simultaneously.
Community-Led Water Innovation: Water solutions based on local observation, kinship, Elders' teachings, and adaptability. Not based on external standards.
Nipiy (Water): Described throughout the book as a living being, relative, teacher, and holder of memory. Caring for water is a spiritual, legal, and ecological responsibility.
Renewable Water Systems: Indigenous-led innovations such as solar-powered water systems aligning sustainability with cultural responsibility.
Water as Life: A central teaching explaining water as spiritual, political, ecological, and ceremonial life-force. Not metaphorical—literally life.
Water Sovereignty: Indigenous-led governance of water systems grounded in rights, responsibilities, ceremony, and community protocols rather than colonial regulation.

Climate Crisis, Environmental Change, and Colonialism

Colonial Water Governance: Systems based on extraction, efficiency, and technocratic planning that sever Indigenous spiritual, linguistic, and governance relationships with water.
Ecological Grief: Grief experienced when water, animals, land, and climate systems are damaged. Tied to cultural loss and spiritual disruption.
Environmental Racism / Structural Exclusion: Policies and systems that restrict Indigenous access to water infrastructure, decision-making, and sovereignty.
Polycrisis: Multiple simultaneous crises—climate change, biodiversity loss, colonialism, cultural disruption—producing compounding impacts on Indigenous communities.

Energy Sovereignty Terminologies

Energy Sovereignty: Indigenous right and responsibility to govern energy systems—solar, wind, wood, and seasonal cycles—according to Indigenous laws and teachings. Appears in multiple chapters, including renewable energy projects.
Indigenous Energy Innovation: Technologies developed or adapted to respect relational ethics, reduce extraction, and align with ceremony and community needs.

Land-Based Energy Knowledge: Traditional teachings that link energy to land rhythms, kinship, intergenerational responsibility, and medicine.

Health, Healing, and Medicine Terminologies

Kinship Healing Systems: Healing that arises from relationships, community involvement, and ceremonial responsibility.

Land-Based Healing: Healing ceremonies, plant medicines, youth–Elder teachings, and cultural practices connected to land, water, and spirituality.

Traditional Medicines and Food Sovereignty: Health systems rooted in local foods, plants, and ceremonial practices disrupted by climate change. Detailed in later chapters.

Language, Culture, and Governance Terminologies

Cultural Camps / Land-Based Camps Spaces where traditional knowledge is transmitted intergenerationally through hands-on practice, ceremony, and collective learning.

Cultural Memory Teachings, stories, and practices carried through generations, often threatened by environmental loss and colonial policy.

Language as Governance: Indigenous languages are ecological memory systems encoding law, responsibility, seasonal knowledge, and water ethics.

Community Governance Terminologies

Community-Led Sustainability. Environmental decision making emerging from Elders, youth, and community members grounded in land and ceremony—not imposed policy.

Intergenerational Responsibility: The commitment to protect land and water for future generations.

Resurgence: Renewal of Indigenous governance, ceremony, language, identity, and political presence.

Critical Research and Methodology Terminologies

Decolonial Sustainability: Sustainability shaped by Indigenous knowledge, ceremony, kinship, and sovereignty—not by colonial environmental frameworks.

Land-Based Research Methodology: Research that emerges from ceremony, Elders' teachings, relational accountability, and land participation—not extractive academic methods.

Reciprocity: Ethical obligation to give back to communities, land, and water.

Relational Governance: Governance grounded in relationships rather than rules or technocratic metrics.

Key Political and Ethical Terms

Indigenous Sovereignty: Self-determination over land, water, governance, and cultural knowledge.

Solidarity: Walking alongside Indigenous communities with humility, responsibility, and long-term commitment.

Treaty Responsibility: Responsibilities between peoples and land, not only legal agreements.

Compilation of References

Absolon, K. (2011). *Kaandossiwin: How we come to know*. Fernwood Publishing.

Anderson, K. (2011). *Life stages and Native women: Memory, teachings, and story medicine*. University of Manitoba Press. DOI: 10.1515/9780887554056

Anderson, K. (2018). *A recognition of being: Reconstructing Native womanhood* (2nd ed.). Women's Press.

Anderson, K., Clow, B., & Haworth-Brockman, M. (2013). *Carrying water: Aboriginal women, water and health*. Atlantic Centre of Excellence for Women's Health.

Anderson, K., Clow, B., & Haworth-Brockman, M. (2019). Carriers of water: Aboriginal women's experiences, relationships, and reflections. *Journal of Indigenous Wellbeing: Te Mauri-Pimatisiwin*, *4*(1), 1–19.

Arsenault, R., Bourassa, C., Diver, S., McGregor, D., & Witham, A. (2019). Including Indigenous knowledge systems in environmental assessments: Restructuring the process. *Global Environmental Politics*, *19*(3), 120–140. DOI: 10.1162/glep_a_00519

Arsenault, R., Diver, S., McGregor, D., Witham, A., & Bourassa, C. (2018). Shifting the framework of Canadian water governance through Indigenous research methods: Acknowledging water as a living entity. *Water (Basel)*, *10*(5), 452. DOI: 10.3390/w10010049

Arsenault, R., Diver, S., McGregor, D., Witham, J., & Bourassa, C. (2018). Contextualizing water justice: Indigenous water governance in settler colonial Canada. *Water (Basel)*, *10*(5), 1–19. PMID: 30079254

Baker, J., Smith, L., & Richmond, C. (2022). Colonial infrastructures and Indigenous water insecurity in Canada. *Global Environmental Change*, *72*, 102–118.

Bang, M., Marin, A., Faber, L., & Suzukovich, E. (2012). Reframing science education from Indigenous worldviews. *Cultural Studies of Science Education*, *7*(2), 535–547.

Bang, M., Marin, A., Faber, L., & Suzukovich, E. (2016). Repatriating Indigenous land-based pedagogies. *Harvard Educational Review*, *86*(1), 1–24.

Borrows, J. (2010). *Canada's Indigenous Constitution*. University of Toronto Press.

Borrows, J. (2020). *Law's Indigenous ethics*. University of Toronto Press.

Borrows, J. (2020). *Law's Indigenous Ethics*. University of Toronto Press.

Chapola, J., Datta, R., Starlight, T., Hurlbert, M., & Poggendorf, S. (2025). Indigenous women-led climate crisis solutions from decolonial feminist perspectives in Western Canada. *Environmental Science & Policy*, *174*, 104272. DOI: 10.1016/j.envsci.2025.104272

Cook, S. (2014). *The dancing plants: Healing medicines and Indigenous knowledge*. University of Manitoba Press.

Corntassel, J. (2012). Re-envisioning resurgence: Indigenous pathways to decolonization and sustainable self-determination. *Decolonization*, *1*(1), 86–101.

Coulthard, G. S. (2014). *Red skin, white masks: Rejecting the colonial politics of recognition*. University of Minnesota Press. DOI: 10.5749/minnesota/9780816679645.001.0001

Craft, A. (2017). *Breathing life into the Stone Fort Treaty: An Anishinabe understanding of treaty one*. UBC Press.

Craft, A. (2021). *Anishinaabe Nibi Inaakonigewin Report: Reflecting the water laws and responsibilities of the Anishinaabe*. University of Manitoba Press.

Craft, A. (2021). *Anishinaabe Nibi inaakonigewin: The legal principles of water*. University of Manitoba Press.

Craft, A., McGregor, D., & McKay, C. (2019). Indigenous legal traditions and water governance. *McGill Law Journal. Revue de Droit de McGill*, *64*(4), 755–790. DOI: 10.7202/1060866ar

Daigle, M. (2018). Resurging through Kishiichiwan: The spatial politics of Indigenous water relations. *Decolonization*, *7*(1), 159–172.

Datta, R. (2015). A relational theoretical framework and meanings of land, water, and air: A decolonial, Indigenous approach to environmental justice. *Journal of Environmental Studies and Sciences*, *5*(3), 1–12.

Datta, R. (2018). Decolonizing both researcher and research and its effectiveness in Indigenous research. *Research Ethics*, *14*(2), 1–24. DOI: 10.1177/1747016117733296

Duran, E. (2006). *Healing the soul wound: Counseling with American Indians and other Native peoples*. Teachers College Press.

Ford, J. D., King, N., Galappaththi, E. K., Pearce, T., McDowell, G., & Harper, S. L. (2020). The resilience of indigenous peoples to environmental change. *One Earth*, *2*(6), 532–543. DOI: 10.1016/j.oneear.2020.05.014

Gone, J. P. (2021). *A complementarity model for Indigenous healing*. In *APA handbook of multicultural psychology* (Vol. 2). American Psychological Association.

Gone, J. P. (2013). Redressing First Nations historical trauma: Theorizing mechanisms for Indigenous culture as mental health treatment. *Transcultural Psychiatry*, *50*(5), 683–706. DOI: 10.1177/1363461513487669 PMID: 23715822

Green, J. (2021). *Indigenous health and the politics of healing*. University of Toronto Press.

Hart, M. A. (2002). *Seeking Mino-Pimatisiwin: An Aboriginal approach to helping*. Fernwood Publishing.

Hart, M. A. (2010). Indigenous worldviews, knowledge, and research: The development of an Indigenous research paradigm. *Journal of Indigenous Voices in Social Work*, *1*(1), 1–16.

Hinton, L., Huss, L., & Roche, G. (Eds.). (2018). *The Routledge handbook of language revitalization*. Routledge. DOI: 10.4324/9781315561271

Hoicka, C. E., Savic, K., & Campney, A. (2022). Indigenous energy sovereignty in Canada: Policy, power, and colonialism. *Energy Research & Social Science*, *89*, 102–112.

Houle, K. (2022). Indigenous ecological grief and climate pandemic realities. *Environmental Humanities*, *14*(1), 35–56.

Indigenous Services Canada. (2024). *Long-term drinking water advisories on public systems on reserves*. Government of Canada. https://www.sac-isc.gc.ca

Kelly, E. N., Schindler, D. W., Hodson, P. V., Short, J. W., Radmanovich, R., & Nielsen, C. C. (2010). Oil sands development contributes elements toxic at low concentrations to the Athabasca River and its tributaries. *Proceedings of the National Academy of Sciences of the United States of America*, *107*(37), 16178–16183. DOI: 10.1073/pnas.1008754107 PMID: 20805486

Kimmerer, R. W. (2013). *Braiding sweetgrass: Indigenous wisdom, scientific knowledge, and the teachings of plants*. Milkweed Editions.

Kovach, M. (2009). *Indigenous methodologies: Characteristics, conversations, and contexts*. University of Toronto Press.

LaDuke, W. (2020). *To be a water protector: The rise of the Wiindigoo slayers*. Fernwood Publishing.

LaDuke, W. (2020). *To Be a Water Protector: The Rise of the Wiindigoo Slayers*. Fernwood Publishing.

Latulippe, N., & Klenk, N. (2020). Making room and moving over: Knowledge co-production, Indigenous knowledge sovereignty and the politics of global environmental change decision-making. *Current Opinion in Environmental Sustainability*, *42*, 7–14. DOI: 10.1016/j.cosust.2019.10.010

Lavallée, L., & Poole, J. (2010). Beyond recovery: Colonization, health and healing for Indigenous people in Canada. *International Journal of Mental Health and Addiction*, *8*(2), 271–281. DOI: 10.1007/s11469-009-9239-8

Levac, L., McMurtry, J. J., & Walters, D. (2023). Energy poverty, equity, and Indigenous-led transitions. *Energy Policy*, *176*, 113–127.

Lewis, M., & Datta, R. (2023). Resurgence and relational governance in Indigenous climate leadership. *Sustainability Science*, *18*(4), 1123–1137.

Linklater, R. (2014). *Decolonizing trauma work: Indigenous stories and strategies*. Fernwood Publishing.

McCarty, T. L., & Lee, T. S. (2014). Critical culturally sustaining/revitalizing pedagogy and Indigenous education sovereignty. *Harvard Educational Review*, *84*(1), 101–124. DOI: 10.17763/haer.84.1.q83746nl5pj34216

McCarty, T. L., & Nicholas, S. E. (2014). Language education, Indigenous revitalization, and building resilient communities. *Annual Review of Applied Linguistics*, *34*, 21–41.

McCarty, T., & Nicholas, S. (2014). Language loss in Indigenous communities and the environmental crises of meaning. *Annual Review of Anthropology*, *43*, 1–17.

McGregor, D. (2012). Traditional knowledge: Considerations for protecting water in Ontario. *International Indigenous Policy Journal*, *3*(3), 1–21. DOI: 10.18584/iipj.2012.3.3.11

McGregor, D. (2014). Traditional knowledge and water governance: The ethic of responsibility. *Alternative*, *10*(5), 493–507. DOI: 10.1177/117718011401000505

McGregor, D. (2021). Indigenous environmental justice and sustainability. *The Canadian Journal of Native Studies*, *41*(1), 85–110.

McGregor, D. (2021). Indigenous water governance and sustainability. In Renzetti, S., & Dupont, D. P. (Eds.), *Water policy and governance in Canada* (pp. 193–214). Springer., DOI: 10.1007/978-3-030-77502-4_9

McGregor, D., Restoule, J. P., & Johnston, R. (2020). *Indigenous research: Theories, practices, and relationships*. Canadian Scholars Press.

McGregor, D., Whitaker, S., & Sritharan, M. (2020). Indigenous environmental justice and sustainability. *Current Opinion in Environmental Sustainability*, *43*, 35–40. DOI: 10.1016/j.cosust.2020.01.007

McIvor, O. (2020). Indigenous language revitalization and applied linguistics: Parallel histories, shared futures? *Annual Review of Applied Linguistics*, *40*, 78–96. DOI: 10.1017/S0267190520000094

Norgaard, K. M. (2019). *Salmon and Acorns Feed Our People*. Rutgers University Press.

Norman, E. S., Bakker, K., & Cook, C. (2023). Water governance and Indigenous rights in settler states. *Water International*, *48*(1), 1–20. DOI: 10.1080/02508060.2022.2147365

O'Neil, A., Sojo, V., Fileborn, B., Scovelle, A. J., & Milner, A. (2018). The# MeToo movement: An opportunity in public health? *Lancet*, *391*(10140), 2587–2589. DOI: 10.1016/S0140-6736(18)30991-7 PMID: 30070210

Phare, M. A. S. (2009). *Denying the source: The crisis of First Nations water rights*. Rocky Mountain Books Ltd.

Schuster, R., Germain, R. R., Bennett, J. R., Reo, N. J., & Arcese, P. (2019). Vertebrate biodiversity on indigenous-managed lands in Australia, Brazil, and Canada equals that in protected areas. *Environmental Science & Policy*, *101*, 1–6. DOI: 10.1016/j.envsci.2019.07.002

Simms, R., Harris, L. M., Joe, N., & Bakker, K. (2016). Navigating the tensions in collaborative watershed governance: Indigenous peoples, settler colonialism, and water justice. *Human Geographies*, *9*(2), 38–55.

Simms, R., Harris, L. M., Joe, N., & Bakker, K. (2019). Navigating the tensions in collaborative watershed governance: Indigenous rights, participation, and authority. *Water Alternatives*, *12*(1), 1–22.

Simpson, L. (2011). *Dancing on our turtle's back: Stories of Nishnaabeg re-creation, resurgence, and a new emergence.* Arbeiter Ring Publishing.

Simpson, L. (2014). Land as pedagogy: Nishnaabeg intelligence and rebellious transformation. *Decolonization, 3*(3), 1–25.

Simpson, L. B. (2017). *As we have always done: Indigenous freedom through radical resistance.* University of Minnesota Press. DOI: 10.5749/j.ctt1pwt77c

Smith, L. T. (2020). The native and the neoliberal down under: Neoliberalism and "endangered authenticities". In *Indigenous experience today* (pp. 333-352). Routledge.

Smylie, J., Williams, L., & Cooper, N. (2009). Culture-based practices in Indigenous maternal and child health. *Canadian Journal of Public Health = Revue Canadienne de Santé Publique, 100*(3), 229–232.

TallBear. K. (2019). *Caretaking relations, not American dreamings.* In *Critically sovereign: Indigenous gender, sexuality, and feminist studies.* Duke University Press.

Todd, Z. (2016). An Indigenous feminist's take on the ontological turn: 'Ontology' is just another word for colonialism. *Journal of Historical Sociology, 29*(1), 4–22. DOI: 10.1111/johs.12124

Truth and Reconciliation Commission of Canada. (2015). *Honouring the truth, reconciling for the future: Summary of the final report of the Truth and Reconciliation Commission of Canada.*

Tuck, E., & Yang, K. W. (2014). R-Words: Refusing research. In *Humanizing research: Decolonizing qualitative inquiry with youth and communities* (pp. 223–248). SAGE.

Tuck, E., & McKenzie, M. (2015). *Place in research: Theory, methodology, and methods.* Routledge.

Tuck, E., McKenzie, M., & McCoy, K. (2014). Land education: Rethinking pedagogies of place from Indigenous, postcolonial, and decolonizing perspectives. *Environmental Education Research, 20*(1), 1–23. DOI: 10.1080/13504622.2013.877708

Tuck, E., & Yang, K. W. (2012). Decolonization is not a metaphor. *Decolonization, 1*(1), 1–40.

Waldram, J. B. (2008). The models and metaphors of healing: Introduction. *Medical Anthropology Quarterly, 22*(3), 243–245.

Whyte, K. (2017). Indigenous climate justice and settler responsibility. *Daedalus, 146*(3), 115–128.

Whyte, K. P. (2017). Indigenous climate change studies: Indigenizing futures, decolonizing the Anthropocene. *English Language Notes*, *55*(1–2), 153–162. DOI: 10.1215/00138282-55.1-2.153

Whyte, K. P. (2018). Indigenous science (fiction) for the Anthropocene. *Environment and Planning. E, Nature and Space*, *1*(1–2), 224–242. DOI: 10.1177/2514848618777621

Whyte, K. P. (2020). Too late for Indigenous climate justice? *Wiley Interdisciplinary Reviews: Climate Change*, *11*(1), 1–8. DOI: 10.1002/wcc.603

Wildcat, D. R. (2009). *Red alert! Saving the planet with Indigenous knowledge.* Fulcrum Publishing.

Wildcat, D., McDonald, J., Irlbacher-Fox, S., & Coulthard, G. (2014). Learning from the land: Indigenous land-based education. *Decolonization*, *3*(3), 1–15.

Wilson, S. (2008). Research is ceremony: Indigenous research methods. Fernwood Publishing. *(Still foundational; appropriate to cite alongside recent literature)*

Wilson, N. J., & Inkster, J. (2018). Respecting water: Indigenous water governance, ontologies, and the politics of kinship. *Environment and Planning. E, Nature and Space*, *1*(1–2), 223–242. DOI: 10.1177/2514848618789378

Wilson, S. (2008). *Research is ceremony: Indigenous research methods*. Fernwood Publishing.

Wilson, S. (2020). Learning from the land: Indigenous paradigms and research relationships. *Journal of Indigenous Studies*, *15*(2), 45–60.

Yates, J. S., Harris, L. M., & Wilson, N. J. (2017). Multiple ontologies of water: Politics, conflict and implications for governance. *Environment and Planning. D, Society & Space*, *35*(5), 797–815. DOI: 10.1177/0263775817700395

Yazzie, E., & Baldy, C. R. (2018). Diné water politics and relational sovereignty. *Wicazo Sa Review*, *33*(1), 6–28.

About the Authors

Ranjan Datta is the Canada Research Chair in Community Disaster Research at Indigenous Studies, Department of Humanities, Mount Royal University, Calgary, Alberta, Canada. Dr. Datta's current research interests include advocating for Indigenous Land-rights, Indigenous community disaster research, community resiliency on climate change, community-based participatory action research, decolonization, and Indigenous reconciliation. Dr. Datta published 95 peer-reviewed publications, four books, and seven edited books on responsibilities on decolonization, cross-cultural perspectives on reconciliation, Indigenous water, Indigenous climate change, anti-racist perspectives on climate change, and environmental sustainability issues. Dr. Datta has developed a strong understanding of decolonial and Indigenist research frameworks in his 17 years conducting research with Indigenous and non-Indigenous communities in Canada, the USA, Africa, Europe, and South Asia. He is strongly committed to and passionate about Indigenous environmental sustainability, reconciliation, environmental management, Indigenous land rights, anti-racist theory and practice, decolonization, social and environmental justice, community gardens, and cross-cultural research methodology and methods. He has worked and advocated for protecting the Indigenous environment, Land, and sustainability, particularly with South Asian and North American Indigenous communities. He is dedicated to building cross-cultural bridges within Canada among Indigenous, immigrant, and refugee communities. Being born and raised as a minority land-based researcher from Bangladesh, I am reminded that working with Indigenous peoples and communities around the world involves a journey of taking responsibility that can be empowering, rewarding, and challenging at the same time. However, having been displaced from my community's Indigenous Land because of our Indigenous identity and culture, I am thankful to Indigenous peoples in Canada who shared their Land-based health knowledge and practice for my growth. Along my land-based decolonization journey, it is vitally important that I take responsibility for building authentic relationships with the Indigenous people, learning North America colonial histories, and being part of the Indigenous struggle.

Kevin Lewis is an *êhiyaw*(Plains Cree) associate professor, researcher and writer at the University of Saskatchewan. Dr. Lewis has worked with higher learning institutions within the Prairie Provinces of Manitoba, Saskatchewan and Alberta in the areas of Cree language development and instructional methodologies. His research interests include language and policy development, second language teaching methodologies, teacher education programming, and environmental education. For the past 21 years, Dr. Lewis has been working with community schools in promoting land and language-based education and is founder of *kâniyâsihk* Culture Camps (www.kaniyasihkculturecamps.com/),a non-profit organization focused on holistic community well-being and co-developer of Land-Based Cree Immersion School *kâ-nêyâsihk mîkiwâhpa*. Dr. Lewis is fromMinistikwanLake Cree Nation in Treaty 6 Territory. Dr. Kevin Lewis, known as wâsakâyâsiw, is a nêhiyaw (Plains Cree) scholar and educator from Ministikwan Lake Cree Nation in Treaty 6 Territory. His research positionality is deeply rooted in revitalizing Indigenous languages and cultural knowledge. As a knowledge and language keeper, he bridges traditional wisdom with academic scholarship in Indigenous community-led water and energy sustainability. Dr. Lewis launched the Indigenous Languages Certificate (ILC) at the University of Saskatchewan, focusing on Cree dialects and addressing the impact of colonization on language loss. His mission, guided by Elders, emphasizes empowering future generations through nêhiyawêwin, fostering resilience, cultural identity, and a stronger connection to Turtle Island's Indigenous heritage.

Margot Hurlbert is a Professor and Canada Research Chair, Tier 1, Climate Change, Energy, and Sustainability Policy of the Johnson Shoyama Graduate School of Public Policy, University of Regina. She explores the gap between what is needed to address climate change and current policy and behaviour. Margot's scholarship concerns climate change adaptation and mitigation, energy, Indigenous peoples, water, droughts, floods, water governance and sustainability, and achieving net zero emissions. Margot has led and participated in many academic and industry funded research projects, serves on the editorial boards of international journals, and is a Senior Research Fellow of the Earth Systems Governance Project. Margot was Coordinating Lead Author of a chapter of the Special Report of the Intergovernmental Panel ("IPCC") on 'Climate Change on Land' (2019) and a Review Editor and Contributing Author for the IPCC's AR6 (WGI and WGII) (2021/2022). She also worked on Future Earth's Earth Commission Working Group on Transformations (2019-2022) and is an expert panel member on 'Adaptation' for the Canadian Climate Institute and on the Research Board of the World Meteorological Organization (WMO). I entered law school because I had a passion for justice. I eagerly embarked on a practice of law, but soon realized courts and the Canadian legal system always arrived at legal justice. But legal justice often isn't the same as, nor does it achieve, substantive justice. As I have sought justice, and continue to do so, I seek a world without oppression and discrimination, where everyone has the same opportunities and life chances. To achieve this the words and knowledge of people that are oppressed and discriminated against need to be heard. I am a supporter, an ally that walks with and beside Indigenous peoples and visible minorities. I am part of the cheering crowd, not standing in front nor hiding behind. We are a team of community-based researchers who engage in ethical Indigenous research. All our work is informed by our longstanding respectful and reciprocal relationships with Indigenous communities. Our main goal is to promote Indigenous Peoples' sovereignty and self-determination through research. Studies argue that self-determination is the key to closing Indigenous inequities (Reading & Wien, 2009). Particularly, this research will help address the water crises within remote Indigenous communities by highlighting community-led water management and protection. The research will benefit the community by enhancing access to safe drinking water while promoting Indigenous-led water governance.

Jebunnessa Chapola is a settler woman of colour, trained as an anti-racist, decolonial, transnational feminist educator, and active community builder. Currently a contract faculty member at Mount Royal University, Calgary, and an SSHRC postdoctoral fellow at the University of Regina, she collaborates with Professor Margot Hurlbert on climate change, energy, and sustainability policy. With over a decade of community activism in Saskatoon and Calgary, Dr. Chapola has been recognized with numerous accolades, including the MOMA Award (2023) and the CBC Future 40 Award (2015), for her contributions to anti-racism and social justice initiatives. Dr. Jebunnessa Chapola's research positionality is anchored in anti-racist, decolonial, and transnational feminist frameworks, emphasizing community-engaged scholarship. As a settler woman of colour, her work challenges systemic inequities by centering marginalized voices and fostering cross-cultural understanding. Her approach integrates academic inquiry with lived experiences, advocating for social justice and policy transformation. Through collaborations in climate change, energy, and sustainability research, she explores equitable solutions that address intersecting issues of race, gender, and colonialism. Her positionality is deeply intertwined with her activism, striving for inclusive knowledge production and meaningful change within both local and transnational contexts.

Michelle Rose Whitstone is an Assistant Professor at Diné College in Tsaile, Arizona, USA, and a PhD candidate in the Department of Educational Administration at the University of Saskatchewan. My research focuses on Pedagogic Theory, Language Education, and Educational Leadership. Currently, She is working on a project aimed at creating a comprehensive framework for effective language revitalization efforts that can be easily understood and implemented by administrators. As a Diné First Nationa, I am passionate about contributing meaningful literature to the field of heritage language advocacy and revitalization, standing shoulder-to-shoulder with countless other language warriors. In my rare moments of downtime, I enjoy cooking meals over a campfire, singing traditional Diné songs (some of which can be found on YouTube), using my auctioneering skills to support communities, or driving down rugged reservation roads with the best travel companion, my mom. I am deeply committed to land-based, language-immersion learning and often remind those around me: *We cannot respect that which we do not understand.* I firmly believe that to truly transform as a human species, we need to adopt a change model rooted in Indigenous process philosophy. When examined deeply, this approach mirrors the adaptive and integrative nature of artificial intelligence. Indigenous process philosophy encourages us to see the world through a critical lens, fostering an understanding that inspires action against unethical and unjust practices.

John Bosco Acharibasam (BA, MA, Ph.D.) is a Postdoctoral Fellow at Mount Royal University whose research focuses on addressing social and environmental justice issues affecting Black, Indigenous, and other marginalized communities. Dr. Acharibasam's research philosophy prioritizes the co-creation of knowledge, centering the voices and worldviews of Indigenous peoples and other marginalized groups. His primary goal is to reduce environmental vulnerabilities and health inequities within these populations. As an immigrant, John is deeply committed to fostering cross-cultural connections among BIPOC communities in Saskatchewan, Canada. His work also explores decolonizing methodologies, climate change, health disparities, and anti-racism initiatives. I am a community-based researcher whose work is deeply rooted in Indigenous decolonial research frameworks. Collaborating with remote Indigenous communities and equity-deserving groups, I approach research with a commitment to reciprocity, respect, and cultural humility. His positionality is shaped by my lived experiences as an immigrant and his engagement with communities that have historically faced marginalization and systemic inequities. I acknowledge the importance of addressing power imbalances in research relationships and strive to uphold ethical practices that honour community sovereignty and self-determination.

Index

C

ceremonial 2, 3, 4, 5, 6, 7, 8, 9, 10, 11, 12, 13, 14, 15, 16, 17, 18, 19, 20, 23, 25, 26, 28, 29, 32, 33, 35, 36, 39, 40, 43, 44, 45, 46, 47, 48, 49, 50, 51, 52, 53, 54, 55, 56, 57, 58, 59, 61, 63, 64, 68, 69, 70, 71, 72, 74, 77, 78, 79, 80, 81, 82, 85, 86, 87, 89, 90, 91, 92, 94, 95, 97, 98, 99, 101, 105, 106, 108, 111, 112, 113, 114, 115, 116, 117, 118, 120, 121, 122, 125, 126, 128, 129, 131, 133, 134, 135, 136, 137, 139, 140, 144, 145, 147, 148, 152, 153, 155, 158, 163, 164, 165, 166, 169, 170, 171, 172, 173, 174, 175, 176, 177, 178, 179, 180, 182, 183, 184, 187, 188, 189, 191, 192, 194, 195, 196, 197, 198, 200, 201, 202, 203, 205, 206

ceremonies 5, 6, 7, 9, 10, 11, 13, 16, 18, 24, 25, 26, 28, 30, 31, 33, 36, 38, 39, 40, 43, 44, 45, 46, 47, 49, 50, 51, 52, 53, 54, 57, 58, 59, 69, 75, 76, 77, 88, 92, 93, 106, 112, 114, 117, 118, 126, 129, 131, 132, 133, 134, 136, 137, 140, 141, 150, 152, 154, 156, 162, 171, 174, 180, 187, 188, 189, 194, 197, 198, 199, 203, 204, 205, 206

ceremony 1, 2, 3, 4, 5, 7, 8, 9, 10, 11, 12, 13, 14, 15, 16, 19, 20, 22, 24, 34, 37, 39, 40, 43, 44, 45, 46, 48, 50, 51, 52, 53, 54, 55, 56, 57, 58, 59, 61, 62, 63, 64, 65, 66, 67, 69, 70, 74, 75, 76, 77, 78, 79, 80, 81, 85, 86, 90, 92, 94, 95, 96, 97, 102, 103, 105, 108, 109, 111, 112, 115, 116, 117, 118, 119, 120, 121, 122, 125, 126, 128, 129, 130, 132, 133, 134, 135, 136, 137, 139, 140, 141, 142, 143, 145, 146, 147, 148, 149, 155, 156, 157, 161, 162, 164, 165, 167, 168, 169, 170, 171, 172, 173, 174, 178, 179, 180, 181, 182, 183, 184, 185, 186, 187, 188, 189, 190, 192, 193, 194, 195, 196, 197, 198, 199, 200, 201, 202, 203, 205, 206

climate 2, 3, 4, 5, 6, 7, 9, 10, 11, 12, 16, 17, 18, 19, 20, 21, 22, 26, 28, 29, 30, 33, 35, 36, 38, 39, 40, 41, 43, 46, 48, 58, 59, 61, 62, 64, 73, 74, 75, 77, 85, 87, 89, 90, 91, 93, 94, 102, 105, 106, 109, 111, 139, 140, 142, 143, 144, 145, 146, 147, 149, 150, 153, 154, 156, 157, 158, 160, 163, 167, 168, 169, 170, 186, 187, 188, 189, 206, 207, 208

collective 7, 8, 9, 11, 12, 13, 20, 25, 30, 32, 44, 45, 47, 48, 49, 50, 52, 53, 56, 59, 66, 70, 77, 78, 86, 88, 96, 105, 108, 111, 116, 118, 120, 126, 129, 133, 136, 141, 142, 145, 158, 161, 162, 163, 164, 167, 168, 170, 172, 173, 175, 181, 182, 184, 187, 188, 189, 190, 192, 195, 196, 197, 198, 199, 200, 201, 204, 205

colonial 1, 2, 4, 5, 6, 9, 10, 11, 12, 13, 16, 17, 18, 19, 20, 21, 24, 25, 28, 29, 30, 31, 34, 37, 38, 41, 43, 44, 45, 46, 47, 57, 58, 59, 61, 67, 70, 72, 82, 85, 86, 87, 88, 95, 96, 97, 98, 99, 100, 101, 103, 108, 109, 110, 111, 112, 121, 129, 141, 143, 145, 146, 147, 149, 150, 157, 162, 163, 167, 168, 169, 170, 171, 172, 173, 179, 184, 188, 189, 190, 191, 194, 205, 206, 207

communities 2, 3, 4, 5, 6, 7, 8, 10, 11, 12, 14, 15, 16, 17, 18, 20, 22, 23, 24, 25, 26, 28, 29, 30, 31, 32, 33, 34, 36, 37, 38, 39, 40, 41, 43, 44, 46, 47, 48, 49, 50, 51, 52, 53, 54, 57, 58, 59, 61, 62, 63, 64, 65, 66, 67, 68, 70, 71, 72, 73, 74, 75, 76, 77, 78, 81, 91, 92, 95, 96, 98, 99, 100, 105, 106, 108, 109, 111, 117, 119, 120, 126, 129, 138, 140, 141, 145, 146, 147, 148, 149, 150, 154, 157, 162, 163, 165, 166, 167, 170, 172, 178, 180, 184, 188, 190, 191, 192, 193, 195, 199, 200, 201, 203, 206

community 2, 3, 4, 5, 6, 7, 8, 9, 10, 11, 13, 14, 15, 16, 17, 18, 19, 23, 25, 28, 29, 30, 31, 33, 36, 37, 38, 39, 40, 46, 47, 50, 51, 52, 53, 55, 57, 58, 59, 61, 62,

63, 64, 65, 66, 67, 70, 71, 72, 73, 76, 78, 81, 82, 85, 88, 89, 90, 92, 93, 94, 95, 97, 101, 102, 103, 105, 106, 108, 109, 110, 112, 113, 114, 118, 119, 120, 121, 126, 128, 131, 134, 136, 137, 139, 141, 143, 144, 145, 147, 148, 149, 150, 153, 154, 155, 156, 157, 161, 162, 163, 164, 165, 166, 168, 170, 171, 174, 176, 177, 179, 180, 181, 182, 183, 184, 187, 188, 189, 190, 191, 192, 195, 198, 199, 204, 205, 206

cultural 1, 2, 3, 4, 5, 6, 7, 8, 9, 10, 11, 12, 13, 14, 15, 16, 17, 18, 19, 21, 23, 24, 25, 26, 28, 29, 30, 31, 32, 33, 34, 35, 36, 37, 38, 39, 40, 43, 44, 45, 46, 47, 48, 49, 50, 51, 52, 53, 55, 57, 58, 59, 62, 66, 69, 70, 71, 72, 73, 74, 75, 76, 85, 87, 88, 89, 90, 91, 92, 93, 95, 96, 97, 98, 99, 100, 101, 102, 103, 109, 110, 111, 112, 116, 118, 119, 126, 128, 136, 137, 141, 145, 146, 147, 149, 152, 153, 155, 157, 162, 164, 167, 168, 169, 170, 171, 172, 173, 175, 187, 188, 189, 190, 194, 196, 199, 204, 205

E

ecological 1, 2, 3, 4, 6, 7, 9, 11, 12, 16, 18, 19, 21, 23, 24, 25, 28, 30, 32, 33, 34, 35, 36, 37, 38, 39, 40, 43, 44, 45, 46, 47, 48, 49, 51, 52, 55, 57, 58, 59, 61, 62, 63, 64, 65, 69, 70, 74, 75, 76, 77, 81, 85, 86, 87, 88, 89, 90, 91, 92, 96, 97, 99, 100, 101, 102, 103, 105, 108, 109, 116, 117, 118, 125, 137, 139, 141, 143, 145, 146, 147, 149, 150, 157, 158, 160, 161, 162, 163, 166, 168, 170, 171, 173, 179, 180, 181, 183, 188, 189, 190, 191, 192, 193, 197, 200, 204, 205, 206, 208
economic 6, 13, 19, 23, 24, 28, 29, 30, 31, 38, 40, 101, 148, 161, 162, 191, 204
emotional 3, 12, 18, 25, 28, 29, 34, 35, 38, 47, 49, 51, 53, 55, 56, 62, 64, 68, 70, 74, 76, 79, 80, 81, 92, 100, 105, 106, 108, 109, 110, 114, 115, 118, 119, 120, 122, 125, 126, 128, 129, 130, 132, 133, 134, 135, 136, 143, 163, 164, 165, 167, 168, 169, 170, 171, 175, 176, 180, 181, 182, 183, 189, 194, 196, 199, 201, 203, 204, 205
energy 1, 2, 3, 4, 5, 6, 7, 9, 11, 12, 13, 14, 15, 16, 17, 18, 19, 21, 22, 35, 112, 114, 129, 130, 131, 166, 172, 175
environmental 1, 2, 3, 4, 5, 6, 9, 10, 11, 12, 13, 14, 15, 16, 17, 18, 19, 20, 21, 22, 23, 24, 25, 26, 28, 29, 32, 33, 34, 35, 36, 37, 38, 39, 40, 41, 43, 44, 45, 46, 47, 48, 49, 50, 51, 52, 53, 57, 58, 59, 61, 62, 63, 64, 65, 66, 67, 69, 70, 72, 74, 75, 76, 77, 78, 81, 85, 86, 87, 88, 89, 90, 91, 92, 93, 94, 97, 101, 105, 111, 115, 118, 120, 121, 122, 125, 141, 142, 143, 144, 149, 155, 156, 159, 162, 163, 166, 168, 172, 173, 177, 178, 179, 181, 185, 186, 187, 188, 189, 190, 192, 193, 194, 195, 197, 198, 199, 200, 204, 205, 206, 207
ethical 1, 3, 4, 5, 6, 7, 8, 9, 10, 12, 13, 15, 16, 18, 20, 29, 30, 35, 36, 40, 44, 57, 59, 72, 81, 82, 85, 87, 89, 91, 92, 94, 96, 100, 101, 103, 105, 116, 139, 140, 141, 142, 143, 145, 146, 147, 149, 150, 161, 162, 163, 164, 173, 174, 175, 177, 178, 179, 183, 184, 188, 189, 194, 195, 196, 197, 198, 199, 200, 201, 202, 203, 204, 205, 206

G

governance 1, 2, 3, 4, 5, 7, 8, 9, 10, 11, 12, 13, 14, 15, 16, 17, 18, 19, 20, 21, 22, 24, 26, 32, 40, 41, 43, 44, 45, 46, 47, 49, 50, 52, 54, 56, 57, 58, 59, 61, 62, 63, 65, 66, 67, 70, 71, 72, 73, 81, 82, 85, 86, 87, 88, 89, 92, 93, 94, 96, 99, 103, 116, 125, 126, 128, 139, 140, 141, 142, 143, 144, 145, 146, 147, 148, 149, 150, 151, 152, 153, 154, 155, 156, 158, 159, 161, 162, 163, 164, 165, 166, 167, 168, 170, 171, 172, 173, 174, 175, 177, 178, 179,

181, 182, 183, 185, 186, 187, 188, 189, 190, 191, 192, 193, 194, 195, 196, 197, 198, 199, 200, 201, 202, 203, 204, 205, 206, 207

H

healing 5, 7, 10, 11, 15, 19, 20, 28, 30, 37, 39, 43, 44, 45, 46, 50, 52, 53, 56, 57, 58, 59, 61, 62, 68, 69, 71, 77, 78, 79, 80, 81, 95, 96, 97, 105, 106, 108, 109, 110, 111, 112, 113, 114, 115, 116, 117, 118, 119, 120, 121, 122, 123, 125, 126, 128, 129, 130, 131, 132, 133, 134, 135, 136, 137, 138, 141, 164, 165, 166, 167, 168, 169, 170, 171, 172, 173, 174, 175, 176, 177, 178, 179, 180, 181, 182, 183, 184, 185, 187, 188, 189, 196, 198, 199, 200, 203, 204, 205, 206

health 13, 18, 19, 31, 32, 33, 34, 35, 47, 48, 59, 62, 63, 64, 65, 66, 71, 74, 75, 77, 78, 81, 83, 85, 89, 91, 101, 105, 108, 109, 110, 111, 114, 115, 119, 120, 121, 123, 125, 126, 134, 137, 138, 161, 162, 164, 165, 166, 167, 168, 169, 170, 171, 173, 175, 176, 177, 178, 182, 184, 188, 190, 191, 204

highlights 23, 28, 30, 38, 39, 44, 49, 50, 54, 57, 65, 67, 68, 69, 71, 73, 77, 78, 79, 81, 85, 87, 90, 92, 94, 101, 103, 112, 115, 119, 125, 133, 148, 161, 162, 165, 167, 168, 171, 172, 178, 180, 182, 191, 198, 199, 200, 203

I

Indigenous 1, 2, 3, 4, 5, 6, 7, 8, 9, 10, 11, 12, 13, 14, 15, 16, 17, 18, 19, 20, 21, 22, 23, 24, 25, 26, 28, 29, 30, 31, 32, 33, 34, 35, 36, 37, 38, 39, 40, 41, 43, 44, 45, 46, 47, 48, 49, 50, 51, 52, 53, 55, 56, 57, 58, 59, 60, 61, 62, 63, 64, 65, 66, 67, 68, 69, 70, 71, 72, 73, 74, 75, 76, 77, 78, 79, 80, 81, 82, 85, 86, 87, 88, 89, 90, 91, 92, 93, 94, 95, 96, 97, 98, 99, 100, 101, 102, 103, 104, 105, 106, 108, 109, 110, 111, 112, 113, 114, 115, 116, 117, 118, 119, 120, 121, 122, 123, 125, 126, 127, 128, 129, 130, 131, 132, 133, 134, 135, 136, 137, 138, 139, 140, 141, 142, 143, 144, 145, 146, 147, 148, 149, 150, 151, 152, 153, 154, 155, 156, 157, 158, 159, 160, 161, 162, 163, 164, 165, 166, 167, 168, 169, 170, 171, 172, 173, 174, 175, 176, 177, 178, 179, 180, 181, 182, 183, 184, 186, 187, 188, 189, 190, 191, 192, 193, 194, 195, 196, 197, 198, 199, 200, 201, 202, 203, 204, 205, 206, 207, 208

industrial 43, 44, 45, 46, 49, 57, 58, 59, 61, 62, 63, 64, 65, 66, 74, 78, 151, 153, 188, 190, 191, 194, 204

intergenerational 1, 3, 5, 6, 7, 10, 11, 12, 13, 14, 15, 23, 24, 25, 26, 28, 32, 33, 35, 38, 39, 40, 43, 44, 47, 50, 52, 58, 59, 61, 62, 63, 72, 77, 85, 88, 89, 92, 93, 94, 95, 96, 98, 99, 100, 103, 105, 106, 108, 110, 111, 112, 114, 117, 118, 119, 121, 122, 125, 126, 128, 133, 134, 135, 136, 139, 143, 146, 148, 156, 158, 162, 163, 165, 167, 168, 169, 170, 172, 173, 177, 184, 188, 189, 190, 191, 193, 195, 197, 198, 200

K

Knowledge 1, 2, 3, 4, 5, 6, 7, 8, 9, 10, 11, 12, 13, 14, 15, 16, 17, 19, 21, 22, 23, 24, 25, 26, 28, 29, 30, 31, 32, 33, 34, 35, 36, 37, 38, 39, 40, 41, 43, 44, 45, 46, 48, 49, 51, 54, 56, 57, 58, 61, 62, 63, 64, 66, 72, 73, 78, 85, 86, 87, 88, 89, 90, 91, 92, 93, 94, 95, 96, 97, 98, 100, 101, 102, 103, 105, 106, 108, 110, 113, 114, 115, 117, 118, 119, 123, 125, 128, 130, 133, 134, 135, 136, 137, 138, 139, 140, 141, 144, 145, 146, 147, 148, 149, 150, 152, 153, 154, 155, 159, 161, 162, 163, 164, 165, 166, 167, 168, 171, 172, 173, 174, 175, 176, 177, 178, 179,

181, 182, 183, 184, 187, 188, 189, 190, 191, 193, 197, 198, 200, 201, 205, 206, 207

L

land 1, 2, 3, 4, 5, 6, 7, 8, 9, 10, 11, 12, 13, 14, 15, 16, 17, 18, 19, 20, 21, 22, 23, 24, 25, 26, 28, 29, 30, 31, 32, 33, 34, 35, 36, 37, 38, 39, 40, 41, 43, 44, 45, 46, 47, 48, 49, 51, 52, 53, 54, 55, 56, 58, 59, 61, 62, 63, 64, 65, 66, 67, 68, 69, 70, 71, 72, 73, 74, 75, 77, 78, 79, 80, 81, 85, 86, 87, 88, 89, 90, 91, 92, 93, 94, 95, 96, 97, 98, 99, 100, 101, 102, 103, 105, 106, 108, 109, 110, 111, 112, 113, 114, 117, 118, 119, 120, 121, 122, 125, 126, 127, 128, 129, 130, 131, 133, 134, 135, 137, 138, 139, 140, 141, 142, 143, 144, 145, 146, 147, 148, 149, 150, 151, 152, 153, 154, 155, 156, 157, 158, 159, 161, 164, 165, 166, 167, 168, 169, 170, 171, 172, 173, 174, 175, 176, 177, 178, 179, 180, 181, 182, 183, 184, 185, 188, 189, 190, 191, 192, 193, 194, 196, 197, 198, 199, 200, 202, 203, 205

land-based 1, 2, 3, 4, 5, 6, 7, 8, 9, 10, 11, 12, 13, 14, 15, 16, 17, 18, 19, 20, 23, 25, 26, 28, 29, 31, 32, 33, 34, 35, 36, 37, 38, 39, 40, 41, 43, 44, 46, 47, 48, 49, 51, 52, 53, 55, 56, 61, 62, 63, 64, 66, 67, 68, 69, 70, 72, 73, 74, 75, 77, 78, 79, 80, 85, 86, 88, 89, 90, 91, 92, 93, 94, 95, 96, 97, 99, 100, 102, 103, 105, 125, 126, 127, 130, 131, 133, 134, 138, 139, 140, 141, 142, 143, 144, 145, 146, 147, 148, 149, 150, 151, 152, 153, 155, 156, 158, 164, 165, 166, 167, 168, 169, 170, 171, 172, 173, 174, 175, 176, 177, 178, 179, 180, 181, 182, 183, 184, 185, 189, 190, 192, 193, 194, 196, 197, 198, 199, 200, 202, 203

language 2, 3, 4, 5, 6, 7, 8, 10, 11, 14, 17, 18, 19, 22, 23, 24, 25, 28, 29, 31, 32, 33, 34, 35, 36, 37, 38, 39, 40, 41, 85, 86, 87, 88, 89, 90, 91, 92, 93, 94, 95, 96, 98, 99, 100, 101, 102, 103, 104, 106, 109, 111, 112, 121, 141, 142, 147, 149, 156, 168, 170, 172, 186, 208

languages 4, 9, 13, 19, 23, 24, 25, 26, 28, 29, 32, 33, 34, 35, 36, 37, 38, 39, 40, 85, 86, 87, 88, 89, 90, 91, 92, 93, 94, 95, 96, 97, 98, 100, 103, 139, 147, 171

linguistic 4, 12, 25, 29, 30, 31, 32, 33, 34, 36, 85, 86, 87, 88, 89, 90, 91, 92, 93, 94, 95, 96, 97, 98, 99, 100, 101, 102, 149

M

medicine 37, 39, 40, 77, 80, 81, 96, 97, 105, 106, 108, 110, 112, 113, 114, 115, 116, 117, 118, 119, 120, 121, 125, 127, 128, 129, 130, 131, 137, 157, 159, 164, 165, 166, 167, 169, 172, 173, 174, 175, 176, 177, 178, 179, 181, 182, 183, 184, 185

P

personal 44, 45, 50, 51, 54, 55, 56, 71, 75, 79, 80, 111, 121, 126, 131, 133, 134, 135, 136, 163, 170, 171, 172, 174, 179, 195, 196, 200, 201, 202, 203

practices 1, 2, 3, 6, 9, 10, 11, 13, 15, 16, 17, 20, 23, 24, 25, 26, 28, 30, 31, 32, 34, 35, 36, 37, 39, 40, 43, 44, 45, 46, 47, 51, 52, 54, 55, 56, 57, 58, 59, 61, 62, 63, 66, 67, 68, 71, 74, 81, 82, 85, 86, 87, 88, 89, 90, 91, 95, 96, 97, 103, 105, 106, 110, 111, 112, 113, 114, 116, 117, 119, 120, 121, 123, 125, 126, 127, 130, 131, 133, 134, 135, 136, 137, 138, 139, 141, 144, 145, 146, 148, 149, 150, 152, 154, 159, 161, 162, 163, 164, 165, 169, 170, 171, 172, 173, 175, 176, 177, 178, 179, 180, 181, 182, 184, 185, 187, 188, 189, 193, 194, 195, 196, 197, 202, 204, 205

Q

quote 30, 34, 35, 37, 46, 50, 51, 53, 56, 63, 65, 66, 67, 68, 69, 70, 71, 72, 73, 74, 75, 76, 77, 78, 79, 80, 81, 87, 88, 90, 96, 97, 98, 101, 102, 109, 110, 111, 112, 113, 116, 118, 119, 120, 128, 129, 131, 132, 133, 134, 135, 137, 198, 201, 202

R

relational 1, 2, 4, 5, 6, 7, 8, 9, 10, 11, 12, 13, 14, 15, 16, 18, 19, 20, 21, 22, 23, 24, 25, 28, 30, 34, 35, 36, 37, 38, 39, 40, 43, 44, 46, 48, 49, 50, 53, 54, 55, 56, 57, 58, 59, 62, 66, 77, 78, 79, 81, 82, 85, 86, 87, 89, 90, 92, 93, 94, 96, 97, 100, 101, 102, 103, 105, 106, 108, 109, 110, 111, 113, 115, 116, 117, 118, 119, 120, 121, 125, 126, 128, 130, 134, 135, 136, 137, 138, 139, 140, 141, 142, 143, 144, 145, 146, 147, 148, 149, 150, 151, 153, 154, 155, 157, 158, 160, 161, 162, 163, 164, 165, 167, 169, 170, 171, 172, 173, 174, 175, 176, 177, 178, 179, 180, 181, 182, 183, 184, 187, 188, 189, 190, 191, 192, 193, 194, 195, 196, 197, 198, 199, 200, 201, 202, 203, 204, 205, 206, 208

relationships 2, 3, 4, 5, 6, 7, 8, 9, 10, 11, 12, 13, 14, 15, 16, 18, 19, 20, 22, 23, 24, 25, 26, 28, 30, 32, 33, 35, 37, 39, 40, 43, 44, 45, 46, 51, 53, 54, 55, 56, 57, 58, 59, 61, 63, 65, 67, 70, 74, 77, 79, 81, 86, 88, 89, 94, 95, 97, 100, 101, 102, 103, 105, 108, 113, 114, 118, 122, 123, 126, 135, 139, 140, 141, 142, 143, 145, 146, 147, 159, 161, 162, 163, 164, 165, 166, 168, 170, 171, 172, 175, 178, 180, 181, 183, 188, 190, 191, 192, 193, 194, 195, 196, 197, 198, 199, 201, 202, 204, 207

revitalization 2, 7, 18, 19, 28, 30, 35, 36, 39, 41, 88, 91, 92, 99, 100, 101, 102, 103, 104, 106

rooted 1, 2, 3, 7, 9, 11, 13, 16, 18, 19, 24, 29, 31, 32, 35, 38, 43, 46, 49, 50, 54, 58, 61, 66, 67, 73, 74, 77, 80, 87, 89, 91, 92, 93, 94, 98, 105, 108, 113, 114, 116, 117, 119, 128, 134, 136, 142, 145, 149, 151, 153, 155, 158, 163, 164, 165, 178, 188, 190, 191, 193, 197, 198, 200

S

sacred 4, 6, 13, 20, 23, 31, 36, 43, 44, 45, 46, 48, 49, 50, 51, 52, 53, 55, 56, 57, 58, 59, 61, 62, 63, 65, 66, 67, 69, 70, 73, 74, 75, 77, 80, 81, 87, 91, 92, 94, 95, 97, 101, 106, 112, 113, 114, 115, 116, 117, 119, 120, 121, 122, 128, 129, 130, 131, 132, 133, 134, 135, 137, 142, 152, 153, 154, 156, 163, 178, 179, 181, 182, 184, 185, 189, 190, 192, 193, 196, 198, 199, 200, 202, 203

seasonal 3, 6, 11, 18, 24, 28, 29, 30, 31, 32, 33, 34, 38, 39, 40, 48, 51, 58, 75, 76, 85, 87, 90, 100, 101, 119, 121, 140, 141, 145, 152, 161, 162, 163, 166, 177, 178, 179, 180, 182, 188, 189, 193, 197

sovereignty 1, 2, 3, 4, 5, 6, 9, 10, 11, 13, 15, 16, 17, 18, 21, 26, 28, 30, 44, 52, 61, 63, 67, 68, 72, 77, 82, 85, 88, 89, 90, 91, 96, 103, 104, 128, 138, 139, 140, 142, 143, 144, 146, 147, 149, 150, 151, 156, 157, 158, 159, 163, 172, 173, 184, 191

spiritual 1, 2, 3, 5, 6, 7, 9, 10, 11, 12, 14, 18, 23, 24, 25, 26, 28, 29, 30, 31, 32, 33, 34, 35, 36, 38, 39, 40, 43, 44, 45, 46, 47, 48, 49, 50, 51, 52, 53, 54, 55, 56, 57, 58, 59, 61, 62, 63, 64, 68, 69, 70, 73, 74, 75, 76, 77, 78, 79, 80, 81, 85, 87, 88, 89, 90, 92, 94, 95, 96, 97, 99, 100, 101, 102, 105, 108, 109, 110, 111, 112, 113, 114, 115, 116, 117, 118, 119, 120, 126, 128, 129, 130, 131, 132, 133, 134, 135, 136, 137, 150, 163, 164, 165, 166, 167, 168, 170, 171, 172, 173, 174, 175, 176,

177, 180, 181, 183, 184, 187, 188,
189, 190, 192, 193, 194, 195, 196,
198, 199, 200, 201, 202, 203, 204, 205
subtheme 36, 46, 47, 48, 49, 50, 51, 52,
54, 63, 64, 65, 66, 67, 68, 69, 70, 71,
72, 73, 74, 75, 76, 77, 78, 79, 80, 81,
98, 100, 101, 102, 109, 110, 111, 112,
129, 131, 132, 133, 135, 165, 166, 167,
168, 169, 171, 172, 173, 174, 175, 176,
178, 179, 180, 182, 183
sustain 4, 9, 17, 31, 32, 33, 35, 40, 43, 55,
77, 86, 89, 92, 103, 120, 134, 140, 143,
146, 161, 162, 170, 173, 177, 179,
181, 187, 188, 190, 194, 196, 200, 204
systems 1, 2, 3, 4, 5, 6, 8, 10, 11, 13, 16,
17, 18, 19, 21, 23, 24, 25, 26, 28, 29,
30, 31, 32, 33, 34, 35, 37, 38, 40, 43,
44, 45, 46, 47, 48, 49, 57, 58, 59, 61,
62, 64, 65, 66, 67, 69, 70, 71, 72, 73,
74, 76, 77, 81, 85, 86, 87, 88, 89, 91,
93, 94, 96, 98, 99, 100, 103, 105, 106,
108, 109, 110, 111, 113, 114, 115,
116, 118, 119, 125, 130, 131, 134,
139, 142, 143, 144, 145, 146, 147,
148, 149, 150, 151, 152, 156, 157,
162, 163, 164, 166, 167, 168, 169,
170, 171, 172, 173, 174, 175, 176,
177, 178, 180, 181, 182, 183, 184,
185, 188, 189, 190, 191, 192, 193,
194, 195, 197, 198, 200, 201, 202,
203, 204, 205, 206, 207

T

teachings 1, 2, 4, 5, 7, 8, 9, 10, 11, 13, 14,
16, 17, 19, 20, 21, 24, 25, 29, 30, 31,
32, 33, 34, 35, 36, 37, 38, 39, 40, 41,
43, 44, 46, 47, 49, 50, 51, 53, 54, 57,
58, 59, 61, 62, 63, 68, 69, 71, 73, 77,
79, 81, 85, 87, 88, 90, 91, 92, 93, 94,
96, 97, 98, 99, 101, 105, 108, 109,
116, 117, 118, 119, 120, 121, 122,
123, 125, 128, 129, 130, 131, 133,
134, 135, 136, 137, 138, 139, 141,
146, 147, 150, 152, 155, 156, 157,
159, 161, 163, 165, 167, 168, 169,
170, 172, 173, 177, 178, 183, 184,
187, 188, 189, 191, 193, 197, 200,
201, 202, 204, 205, 207
theme 29, 37, 38, 39, 46, 48, 49, 50, 51,
52, 54, 55, 56, 61, 62, 63, 64, 65, 66,
67, 68, 69, 70, 71, 72, 73, 74, 75, 78,
79, 80, 86, 88, 89, 93, 94, 95, 96, 98,
99, 100, 101, 108, 109, 111, 112, 113,
114, 115, 116, 117, 118, 119, 121,
122, 129, 130, 131, 132, 133, 134,
135, 137, 140, 164, 168, 170, 172,
174, 175, 177, 178, 179, 180, 181,
182, 190, 192, 193, 195, 196, 197,
198, 199, 200, 202, 203
traditional 1, 4, 5, 6, 9, 11, 12, 16, 17, 20,
22, 23, 24, 26, 28, 29, 30, 31, 32, 34,
35, 36, 38, 39, 40, 47, 49, 56, 58, 62,
63, 64, 67, 71, 74, 78, 79, 88, 89, 90,
91, 92, 93, 94, 96, 97, 99, 101, 105,
106, 109, 110, 111, 114, 115, 118,
119, 130, 132, 140, 162, 166, 168,
169, 170, 175, 176, 177, 180, 181,
182, 183, 203, 207
trauma 25, 68, 69, 81, 95, 96, 100, 105,
106, 110, 111, 112, 116, 117, 118,
120, 121, 123, 125, 126, 128, 129,
132, 134, 136, 137, 138, 163, 169,
170, 171, 172, 173, 176, 179, 180,
181, 184, 189, 192

W

water 1, 2, 3, 4, 5, 6, 7, 8, 9, 10, 11, 12, 13,
14, 15, 16, 17, 18, 19, 20, 21, 22, 23,
24, 25, 26, 28, 29, 30, 31, 32, 33, 34,
35, 36, 37, 38, 39, 40, 41, 43, 44, 45,
46, 47, 48, 49, 50, 51, 52, 53, 54, 55,
56, 57, 58, 59, 60, 61, 62, 63, 64, 65,
66, 67, 68, 69, 70, 71, 72, 73, 74, 75,
76, 77, 78, 79, 80, 81, 82, 85, 86, 87,
88, 89, 90, 91, 92, 93, 103, 105, 106,
108, 109, 111, 112, 113, 114, 115,
116, 117, 118, 119, 120, 121, 122,
125, 126, 128, 129, 130, 131, 132,
133, 134, 135, 136, 137, 138, 139,
140, 141, 142, 143, 144, 145, 146,
147, 148, 149, 150, 151, 152, 153,
154, 155, 156, 157, 158, 159, 161,

162, 163, 164, 176, 180, 184, 185, 186, 187, 188, 189, 190, 191, 192, 193, 194, 195, 196, 197, 198, 199, 200, 201, 202, 203, 204, 205, 206, 207
water-based 24, 30, 31, 49, 68, 81, 125, 126, 128, 130, 131, 132, 134, 135, 136, 193

Y

youth 2, 3, 5, 7, 10, 11, 14, 17, 18, 19, 23, 25, 28, 29, 30, 31, 37, 38, 39, 40, 44, 50, 52, 53, 58, 66, 93, 94, 95, 98, 99, 109, 110, 111, 114, 128, 133, 136, 138, 141, 148, 149, 152, 153, 154, 155, 156, 168, 169, 170, 184, 185, 195, 198, 199, 205

www.ingramcontent.com/pod-product-compliance
Lightning Source LLC
Chambersburg PA
CBHW062212220526
45471CB00009B/3175